# Human Endocrinology

# Human Endocrinology

**Paul R. Gard**

Department of Pharmacy,
University of Brighton, UK

UK   Taylor & Francis Ltd, 1 Gunpowder Square, London EC4A 3DE
USA  Taylor & Francis Inc., 1900 Frost Road, Suite 101, Bristol, PA 19007–1598

**British Library Cataloguing–in–Publication Data**

A catalogue record for this book is available from the British Library
ISBN 0-7484-0655-7

**Library of Congress Cataloging–in–Publication Data are available**

Cover design by Jim Wilkie

Typeset in Melior by Keyword Typesetting Services Ltd, Wallington, UK

Printed by T.J. International, Padstow, UK

# Contents

# General Preface to the Series

The curriculum for higher education now presents most degree programmes as a collection of discrete packages or modules. The modules stand alone but, as a set, comprise a general programme of study. Usually around half of the modules taken by the undergraduate are compulsory and count as a core curriculum for the final degree. The arrangement has the advantage of flexibility. The range of options over and above the core curriculum allows the student to choose the best programme for his or her future.

Usually, the subject of the core curriculum, for example biochemistry, has a general textbook that covers the material at length. Smaller specialist volumes deal in depth with particular topics, for example photosynthesis or muscle contraction. The optional subjects in a modular system, however, are too many for the student to buy the general textbook for each and the small in-depth titles generally do not cover sufficient material. The new series *Modules in Life Sciences* provides a selection of texts which can be used at the undergraduate level for subjects optional to the main programme of study. Each volume aims to cover the material at a depth suitable to a particular year of undergraduate study with an amount appropriate to a module, usually around one quarter of the undergraduate year. The life sciences was chosen as the general subject area since it is here, more than most, that individual topics proliferate. For example, a student of biochemistry may take optional modules in physiology, microbiology, medical pathology and even mathematics.

Suggestions for new modules and comments on the present volumes will always be welcomed and should be addressed to the series editor.

*John Wrigglesworth, Series Editor*
King's College London

# 1 Evolution of the Endocrine System and Mechanisms of Hormone Action

## 1.1 Introduction

The body is made up of millions of cells, most of which have a specialized function. For all of these cells to function successfully as a unit there must be some form of communication between them; by considering the following, the development of various forms of intercellular communication can be appreciated.

## 1.2 Evolution of the endocrine system

As shown in Figure 1.1, the simplest form of intercellular communication is for cell A to release a chemical substance into the extracellular fluid. Some of the chemical will arrive at cell B and will influence the activity of that cell, hence the message will have been transmitted. The major problems with this form of communication are that it is non-specific, as many other cells may also receive the chemical message, and that a large amount of transmitter substance must be released in order to ensure that the desired cell receives the message.

If the two cells are further apart, the amount of transmitter substance that must be secreted increases, hence there are greater problems of non-specificity of action, that is, a greater number of cells receive the message that is only intended for cell B. This can be overcome by the development of outgrowths from cell A towards the target cell. The cell may then become further specialized by only releasing transmitter from one area of its surface. This specialization is the beginning of a nerve cell. Nerve cells are very specialized in that they only communicate with specific cells and they only require the release of very small amounts of transmitter.

The drawback of this neuronal form of intercellular communication is that each cell can only communicate directly with a limited number of other cells. The organism therefore requires a method by which a single cell can communicate with a larger number of widely distributed cells. The communication system developed to fulfil that role utilizes a transport system already present within the organism: the blood. Using this second method of intercellular communication less transmitter is required than in the case where transmitter is released into the general extracellular fluid but a large number of cells can be reached by the chemical transmitter substance. There are, however, still some drawbacks in that use of

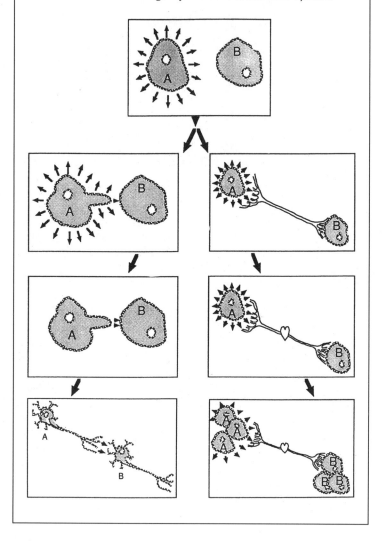

Figure 1.1
A schematic representation of the evolution and development of the endocrine and nervous systems showing the initial communication between cell A and cell B by release of a chemical transmitter (top) and the divergent development of the nervous system (left) and endocrine system (right) to produce means of rapid, specific communication between individual cells (nervous system) and more prolonged, diffuse communication between groups of cells (endocrine system)

the vascular system for transport limits the molecular weight of the transmitter substance because molecules of molecular weight greater than 70000 cannot leave the blood vessels. In addition, transmitter substances can only enter or leave the vascular system via the capillaries; there is no exchange from blood vessels such as arteries, arterioles or veins, hence the communicating cells must be associated with capillary networks. Such direct links between capillary beds are called portal systems; they do exist, but they are very rare. More usually capillary beds are linked via the general circulation in the sequence capillary–venule–vein–heart–artery–arteriole–capillary. Thus if a transmitter substance enters a capillary, within the venous system it will be repeatedly diluted by the arrival of blood from other areas of the body and within the arterial system it will be constantly fractionated by blood leaving to other areas of the body. Only a small proportion of the transmitter leaving cell A thus ever reaches cell B. A large amount of transmitter substance must therefore be secreted initially, and, because of this, the transmitter will reach many other cells in addition to its target cell.

These problems have been overcome in the following ways:

(a) All cells producing the transmitter substance 'clump' together in order to increase the amount of chemical transmitter substance secreted. Such a 'clump' of secretory cells is called a gland.

(b) The chemical transmitter substances are very potent chemicals, so that only very low concentrations are required to produce a response in the target tissue.

(c) The target cells develop receptors to the transmitter substance which are not possessed by other cells; hence only they respond to the transmitter.

In essence, this is a description of the endocrine system. A definition of the endocrine system would be: A system in which a group of secretory cells (a gland) secretes a potent chemical transmitter substance (a hormone) into the blood. The transmitter is then carried by the blood to the target cells where a response is elicited. Endocrine glands are ductless glands, and their hormone transmitters are always secreted into the blood.

The activity of the endocrine system complements that of the nervous system. The nervous system allows for very rapid, but short-lived communication between specific, sometimes individual, cells. The endocrine system allows for a slower, but more prolonged communication between large numbers of cells at many different sites of the body. The endocrine system is therefore the most important factor in the control of the basal processes of the individual, for example metabolism, growth and reproduction.

## 1.3 Anatomy of the endocrine system

Before considering the ways in which the endocrine system influences metabolism, growth, reproduction, etc., it is valuable to consider some basic features of the endocrine system so that similarities and differences can be appreciated and some general principles understood. There are six major glands of the endocrine system, although there are other ones. The anatomy of the endocrine system is given in Figure 1.2. The hypothalamus and pituitary

THE ENDOCRINE SYSTEM
A system in which a group of secretory cells (a gland) secretes a potent chemical transmitter substance (a hormone) into the blood. The transmitter is then carried by the blood to the target cells where a response is elicited.

The activity of the endocrine system complements that of the nervous system. The nervous system allows for rapid, short-lived communication between individual cells; the endocrine system allows for a slower, prolonged communication between large numbers of cells.

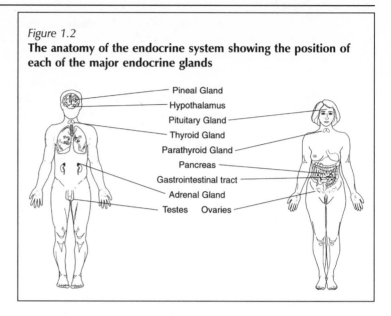

*Figure 1.2*
**The anatomy of the endocrine system showing the position of each of the major endocrine glands**

Pineal Gland
Hypothalamus
Pituitary Gland
Thyroid Gland
Parathyroid Gland
Pancreas
Gastrointestinal tract
Adrenal Gland
Testes     Ovaries

gland, which are associated with the brain, play mainly a regulatory role; it is these glands which control the activities of many of the other endocrine glands, although the pituitary gland does have an endocrine function in its own right. The regulatory role of these structures is emphasized by the fact that it is often said that the pituitary gland is the 'leader of the endocrine orchestra' and that the hypothalamus is the 'conductor'.

Although at first apparently complicated, the naming of hormones is usually logical, and reflects their actions (or origins). Thus a hormone which is secreted by the hypothalamus and acts on the pituitary gland, causing the pituitary to release a second hormone, is called a releasing hormone[1], although in some cases the hypothalamic hormone inhibits pituitary secretion and is thus called a release inhibiting hormone. A hormone which is secreted by the pituitary gland and acts on another endocrine gland to cause the release of a third hormone is called a stimulating or trophic hormone (Figure 1.3).

## 1.4 Chemistry of hormones

One group of hormones is predominantly composed of amino acids, although there are some glycosated moieties; these are the peptide hormones.

Chemically there are two major classes of hormone. The first class is predominantly composed of amino acids, although there are some glycosylated moieties; these will be referred to as the peptide hormones. Many of these hormones are initially synthesized and stored as larger inactive molecules; they are then cleaved to release the active hormone. The second class of hormone is the steroid

[1]By convention if the chemical structure of a releasing hormone has not been identified it is called a releasing factor; however this convention is often ignored, and nomenclature has not kept pace with recent advances.

*Figure 1.3*
**The nomenclature of the endocrine system showing how the name of the individual hormone provides information about the action and origin of that hormone, and its position in the endocrine hierarchy**

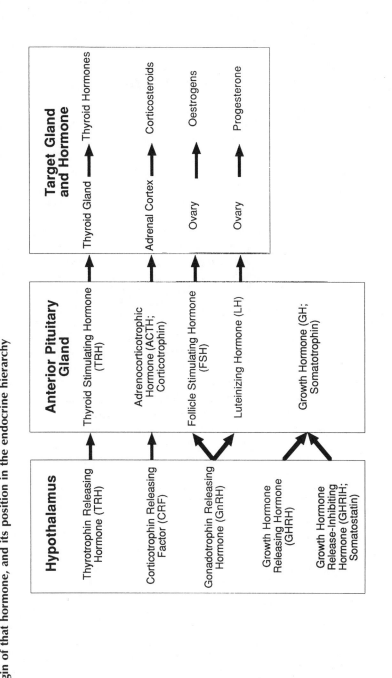

hormones. All of these have a specific four-ringed structure called the cyclopentenoperhydrophenanthrene nucleus. These different classes of hormone have different chemical and biological properties, therefore a knowledge of a hormone's structure can allow some prediction of its mechanism of action (see later). The chemical structures of examples of a peptide and a steroid hormone are presented in Figure 1.4.

## 1.5 Mechanisms of hormone action

The steroid hormones all have a specific four-ringed structure called the cyclopentenoper-hydrophenanthrene nucleus.

In order for a hormone to produce an effect on its target tissue it must be able to influence the activity of the individual cells within that tissue. As is the case for neurotransmitters, mediators of the inflammatory process and most drugs, hormones achieve this by interaction with specific receptors. The way in which a hormone interacts with a receptor to alter cellular activity can be described using the analogy of a lock and key (Figure 1.5). There are many hundreds of different types of receptors that interact with substances such as hormones, neurotransmitters or growth factors, but most cells possess less than ten different types; each cell may possess several thousand of each of the receptor subtypes. Individual receptors will only interact with, and respond to, specific hormones, thus using the analogy of a lock which can only be opened using the appropriate key. Other keys, that is, other hormones or biochemicals, are usually unable to fit into that receptor, or if they can, they are unable to open the lock, that is, produce a response. It can therefore be seen that each hormone must interact with its own specific receptor, and that a cell must possess that receptor if it is to be able to respond to that particular hormone. Furthermore it follows that the hormone must come into direct contact with its receptor to produce a response; hormones cannot act at a distance. The analogy can be carried further because like a lock and key, at a molecular level the hormone and the receptor must have complementary shapes, that is, the hormone must fit the receptor. A slight change in the molecular structure of either the hormone or the receptor may render any interaction impossible. Knowledge of the nature of hormone–receptor interactions has enabled the development of many drugs that are able to modify hormonal activity. Such drugs are chemicals with shapes similar to those of the endogenous hormones, thus some of these drugs are able to bind to the hormone receptor and induce a response similar to that induced by the hormone; these drugs are called agonists. Antagonists are drugs which interact with the hormone receptor but do not produce any effect other than to prevent access of the hormone to the receptor. These antagonists therefore reduce the effect of the hormone in question. Many endocrine disorders that are treated by the use of such hormone analogues, be they agonists or antagonists, are discussed in the following chapters.

*Figure 1.4*
**The chemical structures of oestradiol and gastrin as examples of the steroid and peptide groups of hormones, respectively. The detailed structure is only given for five of the 17 amino acids of gastrin to allow comparison of the relative sizes of the molecules**

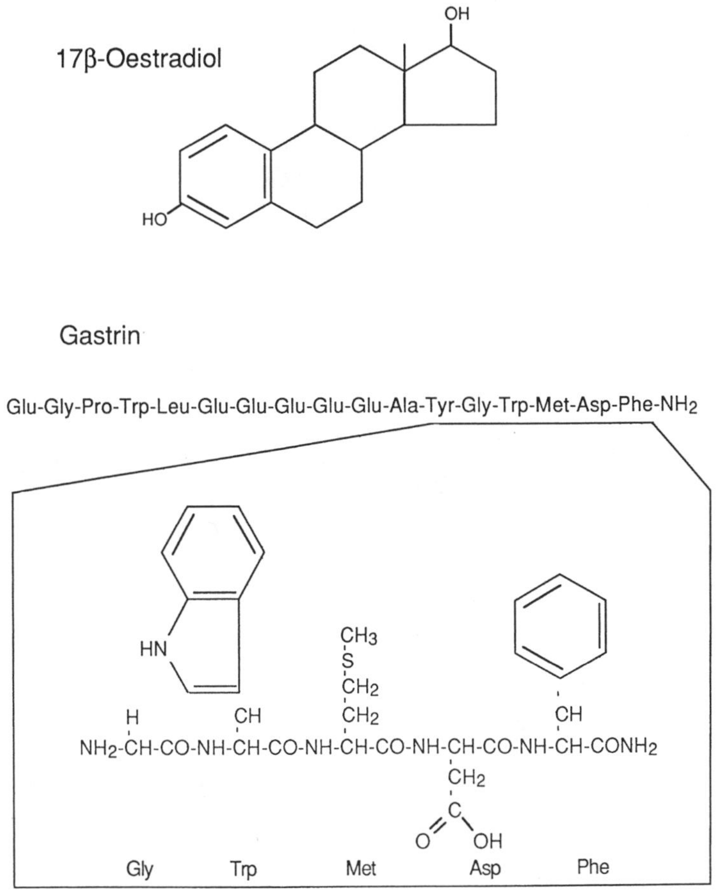
17β-Oestradiol
OH
HO
Gastrin
Glu-Gly-Pro-Trp-Leu-Glu-Glu-Glu-Glu-Glu-Ala-Tyr-Gly-Trp-Met-Asp-Phe-NH2
HN
CH3
S
CH2
CH2
H
CH
CH
NH2-CH-CO-NH-CH-CO-NH-CH-CO-NH-CH-CO-NH-CH-CONH2
CH2
C
O
OH
Gly
Trp
Met
Asp
Phe

All cells have the ability to synthesize all receptors; however variations in the extent of gene expression means that an individual cell is unlikely to synthesize receptors for more than three or four different hormones.

There are many different hormones, and therefore as many different receptors. These receptors are proteins which are synthesized by the cell in the same way as any other proteins are synthesized. The synthesis of these receptor proteins is under genetic control, thus all cells have the ability to synthesize all receptors; however variations in the extent of gene expression means that an individual cell (or tissue) is unlikely to synthesize receptors for more than three or four different hormones. It can therefore be seen why hormones are only able to produce effects on selected target tissues. It is also possible to divide the receptors into different subtypes. The first

Figure 1.5

**A schematic representation of the lock and key hypothesis of hormone–receptor interactions showing how a hormone must have a complementary chemical configuration to that of the receptor in order to produce a response. Hormones or drugs with different structures are unable to produce a response, but may occupy the receptor, thus preventing access of hormone A and hence reducing its effects**

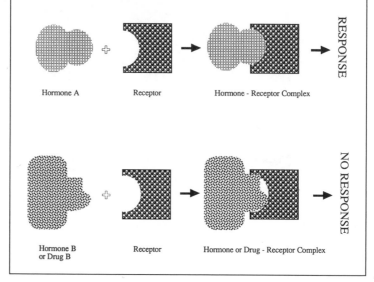

| Hormone A | Receptor | Hormone - Receptor Complex | RESPONSE |

| Hormone B or Drug B | Receptor | Hormone or Drug - Receptor Complex | NO RESPONSE |

receptor classification is based on the site of the receptor within the cell. Some receptors are located within the cell cytoplasm or within the nucleus. In order for the hormone to interact with these receptors it must first cross the lipid cell membrane. Lipid soluble hormones such as the steroid hormones are able to cross the cell membrane readily, thus they are able to interact with intracellular receptors. In the absence of specific membrane transport mechanisms, however, water soluble hormones such as the peptide hormones are unable to cross the cell membrane and thus unable to interact with intracellular receptors. In the case of these hormones the receptors are held within the cell membrane, with that part of the receptor which interacts with the hormone being outside the cell. These hormones are therefore able to interact with the receptor without the necessity of ever crossing the cell membrane or entering the cell. These receptors traverse the cell membrane and therefore have some parts in direct contact with the cytoplasm. It is because these receptors span the cell membrane that the interaction of a hormone with a receptor outside the cell is able to produce an effect inside the cell. Membrane bound receptors are the largest group of receptors and will therefore be the first to be considered in more detail.

### 1.5.1 *Membrane bound receptors*

Within the group of receptors classified as membrane bound receptors there are different subgroups of receptors each of which utilize a different mechanism to produce a change in the intracellular environment and ultimately cellular activity. The two subgroups of membrane bound hormone receptors are G-protein linked receptors and tyrosine kinase linked receptors.

### G-protein linked receptors

A common feature of this group of hormone receptors is that they all utilize a similar protein, called G-protein as a consequence of its guanasine groups. Another common feature of these receptors is their similar structure. These receptors are all long chain proteins comprised of approximately four to five hundred amino acids which are folded in a 'concertina-like' fashion such that they cross the cell membrane seven times (Figure 1.6). These receptors are said to have 'seven transmembrane domains'. The amino end of the protein remains extracellular while the carboxylic end, which is intracellular, interacts with the G-protein which is bound to the intracellular surface of the cell membrane. When the appropriate hormone interacts with the extracellular, amino end of the receptor protein a conformational change in the receptor protein takes place such that the intracellular end stimulates the G-protein. In this way the hormone is able to induce a change inside the target cell without itself having to enter the cell.

G-proteins exist in two states, an inactive state in which the G-protein is bound to guanosine diphosphate (GDP) and an active state in which it is bound to guanosine triphosphate (GTP). The G-protein itself, however, possesses intrinsic GTP-ase activity so it converts the GTP to GDP; thus it normally returns itself to the inactive state. When in the active state, the G-protein is able to influence the activity of some membrane bound enzymes.

Cyclic AMP acts within the cell in which it is produced; it is an example of a second messenger.

The most important membrane bound enzyme with which G-proteins interact is adenyl cyclase. One of two types of G-protein may be linked to the adenyl cyclase: $G_s$, which stimulates the enzyme and $G_i$ which is inhibitory. Adenyl cyclase catalyzes the conversion of adenosine triphosphate (ATP) to cyclic adenosine monophosphate (cyclic AMP or cAMP). The cAMP is then able to act within the cell in which it is produced, but because of its hydrophilic nature it is unable to cross the lipid cell membrane and leave that cell. Cyclic AMP is an example of a second messenger. The first messenger is the hormone; however because it is unable to cross the cell membrane it requires the use of a second messenger to influence processes within the cell. Once produced, the cAMP activates enzymes called cAMP dependent protein kinases. These enzymes are able to alter the activity of key regulatory proteins, such as enzymes, by their phosphorylation. Dependent upon the enzyme concerned the phosphorylation may result in either activation or inhibition of that enzyme. In this way, the interaction of a hormone

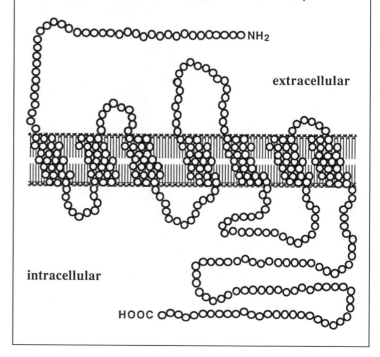

Figure 1.6
**An illustration of a seven transmembrane domain receptor with each circle representing a single amino acid of the receptor protein. Interaction of a hormone with the extracellular portion of the receptor induces a conformational change in the intracellular portion and modification of intracellular processes**

$NH_2$

extracellular

intracellular

HOOC

with its receptor may result in either an increase or a decrease in the metabolic processes of the target cell or tissue. After it has exerted its effect cyclic AMP is destroyed by the enzyme phosphodiesterase; some drugs (e.g. caffeine) inhibit this enzyme and hence prolong the activity of the cyclic AMP. Phosphodiesterase inhibitors may thus potentiate the action of a hormone, or in some cases mimic the hormone's actions (see Box 1.1).

Another type of G-protein called $G_q$ stimulates a different membrane bound enzyme called phospholipase C. This enzyme catalyzes the hydrolysis of phosphoinositides, which are phospholipids normally found within the cell membrane, to produce inositol triphosphate ($IP_3$) and diacylglycerol (DAG), both of which act as second messengers. Like cAMP, these two second messengers influence the activity of intracellular enzyme by causing their phosphorylation: $IP_3$ acts by releasing calcium which in turn activates calcium and calmodulin dependent protein kinases and DAG acts by stimulation of a type of enzyme called protein kinase C. It is interesting to note that DAG, unlike cAMP and $IP_3$, is lipid soluble and is therefore able to leave the cell in which it was produced (Figure 1.7).

DAG, unlike cAMP and $IP_3$, is lipid soluble and is therefore able to cross the cell membrane.

*Box 1.1*   **G-proteins, second messengers and amplification**

As well as allowing water soluble hormones to influence the activity of a cell without having to cross the cell membrane and enter the cell, the G-proteins and second messenger arrangement also allows amplification of the hormonal message. When the hormone interacts with the receptor, it induces a conformational change in that receptor which causes activation of an adjacent G-protein. A single hormone–receptor interaction can lead to the activation of many G-proteins. These G-proteins, in turn, activate associated membrane bound enzymes, such as adenyl cyclase, and because the G-protein may remain in its active form for several seconds, the adenyl cyclase is able to catalyze the production of many molecules of cyclic AMP.

This amplification process is so efficient that a single molecule of hormone can result in the production of 1000 molecules of second messenger.

It is because of this amplification process that a small amount of releasing hormone, produced by a discrete nucleus of cells in the brain, can cause the secretion of a much greater amount of stimulating hormone by the anterior pituitary gland (see Chapter 2). The stimulating hormone is then able to induce the secretion of significant amounts of hormone by glands such as the adrenal cortex. Thus a small change in the activity or function of a small number of cells high in the cascade process can cause immense changes in the activity of the endocrine system.

A fourth type of G-protein, $G_0$, interacts directly with ion channels. Ion channels are composed of proteins that span the cell membrane and are arranged in such a manner that they form a hollow protein cylinder through which ions may pass between the cytoplasm and the extracellular fluid (Figure 1.8). Typically the pores are not constantly open, but rather they open or close in response to various stimuli, for example the interaction of a hormone with a G-protein linked receptor. By influencing the movement of ions, for example calcium, the hormone can influence the cellular activity.

*Figure 1.7*

**An illustration of the inositol second messenger system of a G-protein linked receptor in which interaction of a hormone with the receptor results in activation of the G-protein which, in turn, causes stimulation of phospholipase C and conversion of phosphoinositol to inositol triphosphate and diacylglycerol which act as second messengers**

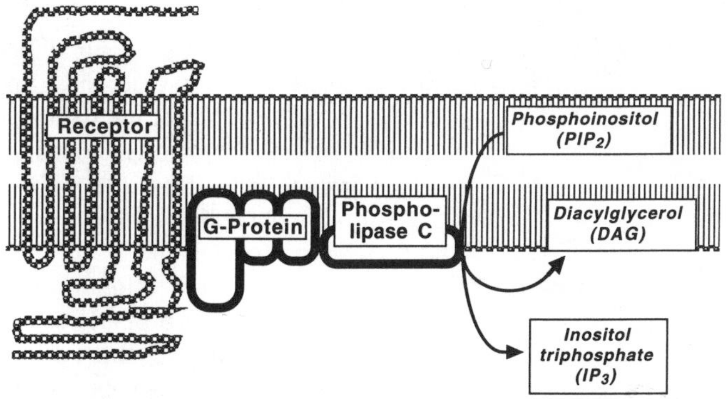

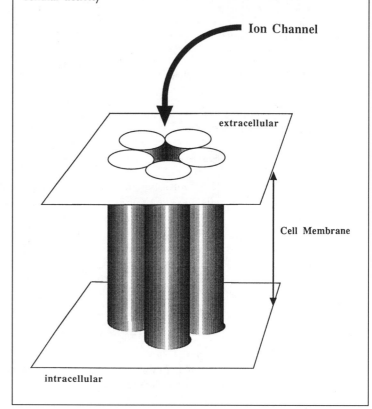

**Figure 1.8**
**A schematic representation of an ion channel in which protein 'cylinders' join together to form a pore through which ions enter or leave a cell. Some hormones act by causing changes in conformation of the protein 'cylinders' thus either opening or closing the ion channels which results in modification of the cellular activity**

Ion Channel

extracellular

Cell Membrane

intracellular

## *Tyrosine kinase linked receptors*

Tyrosine kinase receptors generally consist of a single protein chain which spans the cell membrane. The intracellular portions of these receptors possess the ability to phosphorylate proteins, but they can only do so at the tyrosine residues of the protein. Thus the binding of a hormone with the extraceullar portion of the receptor stimulates protein phosphorylation which may alter the activity of certain intracellular enzymes, or, in this specific case, may influence cell transcription factors thus influencing cell differentiation and replication. One important property of these tyrosine kinase receptors is that they induce the phosphorylation of the tyrosine residues of the receptor protein itself. This has the effect of directly amplifying the activity of the stimulated receptor.

### 1.5.2 *Intracellular receptors*

The other major group of hormone receptors is the intracellular receptors, for example the steroid hormone receptors. For interaction with these receptors the hormones must be able to cross the cell membrane; in most cases this is by simple diffusion of the lipid soluble hormone, but in some cases there is evidence of a carrier mediated transport process. Once in the cell the hormone binds to its receptor, which in some cases normally resides within the cytoplasm and only migrates to the nucleus when bound to the appropriate hormone, or in other cases normally resides within the nucleus. Once in the nucleus, the hormone–receptor complex binds to a specific region of the DNA called the hormone response element (Figure 1.9). The effect of this is to influence processes such as gene expression and RNA transcription. Thus these hormones can cause an increase or decrease in the rate of synthesis of a particular protein, or may initiate the synthesis of a novel protein. It can therefore be seen how hormones such as steroids can produce long lasting effects as the proteins that are produced may be metabolically important enzymes, skeletal proteins (as evidenced by the anatomical changes that are induced by the sex hormones at puberty), or receptors for other hormones or transmitters.

By influencing protein synthesis, steroid hormones produce long-lasting effects, for example changes in anatomy.

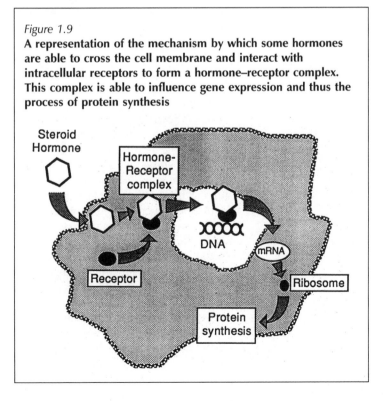

*Figure 1.9*
**A representation of the mechanism by which some hormones are able to cross the cell membrane and interact with intracellular receptors to form a hormone–receptor complex. This complex is able to influence gene expression and thus the process of protein synthesis**

### 1.5.3  *Guanyl cyclase receptors*

The final type of hormone receptors are the guanyl cyclase receptors. These are single strands of protein which may be membrane bound or may be intracellular. In a manner similar to the effects of the seven transmembrane domain receptors on adenyl cyclase, stimulation of these receptors results in the conversion of guanosine triphosphate (GTP) to cyclic guanosine monophosphate (cGMP). The cGMP is able to phosphorylate various intracellular proteins and therefore influence cellular activity. It has been hypothesized that cGMP generally has actions opposite to those produced by cAMP. Thus whereas cAMP may induce smooth muscle contraction, cGMP induces relaxation of that muscle. This is called the 'yin yang' hypothesis; such a relationship has however only been identified in a small number of target tissues.

> cGMP generally has actions opposite to those produced by cAMP: the 'yin yang' hypothesis.

## 1.6 Variations in receptor distribution and population

As already described, some intracellular receptors may move between the cytoplasm and the nucleus, dependent upon their state of activation. Similarly activation of a steroid receptor may either increase or, more commonly, decrease the synthesis of that or another receptor. Thus it can be seen that a given receptor population is not constant, and that the activity of the endocrine system may be influenced not only by the concentrations of circulating hormone, but also by the state of the receptor population.

In the case of the membrane bound receptors it has been shown that these receptors are able to migrate across the cell surface and that they may congregate in clusters of receptors of a similar type. It has also been observed that sections of cell membrane containing hormone receptors may invaginate to produce vesicles. This may occur following binding of the hormone to the receptor where it is a mechanism by which the hormone is rapidly dissociated from the receptor. The hormone may then be broken down by intracellular enzymes while the receptor is recirculated to the membrane ready for restimulation. However it may also occur with unbound receptors. It has been postulated that at times of excessive receptor stimulation the cell membrane invaginates, taking with it receptors, and therefore reducing the extent of the response to the circulating hormone concentration (Figure 1.10). While in the vesicles the receptors may be susceptible to breakdown by enzymes, thus there may be a reduction in the total number of receptors available within that cell. This process is sometimes called receptor down regulation, and is a common response of a receptor population to continued, excessive stimulation.

In some cases, the responses to the circulating hormone may decrease without there being any change in the number of receptors available. This is called receptor desensitization. There are various mechanisms by which this may be achieved including a change in

Figure 1.10

**An illustration of the process of receptor 'down regulation' in which the membrane bound receptors are rendered inaccessible to hormones by invagination of sections of the membrane to form vacuoles. The receptors may later be made accessible by recycling**

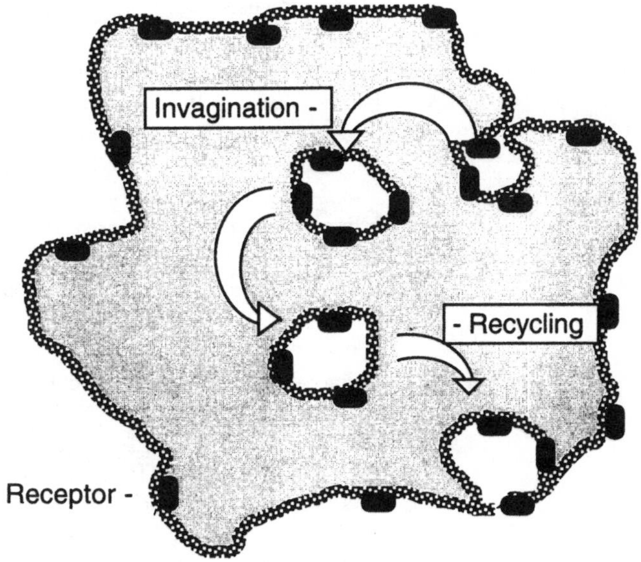

Invagination -

- Recycling

Receptor -

the affinity of the receptor for its hormone, which may indicate a slight change in the structure of the receptor or a decrease in the ability of the receptor to cause the synthesis of its second messenger. In the opposite situation, it has been seen that a prolonged reduction in the concentration of circulating hormone may result in an increase in the expression of the appropriate receptor. This is called receptor up regulation.

The importance of an appreciation of the possibility of a change in receptor activity is that there are some endocrine disorders where there are clinical symptoms of either hormone excess or hormone deficiency in the presence of normal circulating concentrations of that hormone, for example nephrogenic diabetes insipidus (Chapter 3) and some cases of hyperthyroidism (Chapter 4). In these cases the cause of the disorder lies at the receptor level rather than at the glandular level.

# 1.7 Prediction of hormone properties

Earlier in this chapter it was stated that a knowledge of the chemical structure of a hormone allowed some predictions of its biological properties (Table 1.1): peptide hormones are usually stored within glands in the form of an inactive precursor, hence there can be a rapid secretion of the hormone shortly after receipt of the appropriate stimulus by the gland. Peptide hormones are water soluble; they are therefore unable to cross the lipid cell membrane. Because of this they require membrane bound receptors, the stimulation of which usually results in the synthesis of a second messenger which produces effects that are generally rapid in onset. Because these hormones are protein in nature, they are rapidly metabolized and their duration of their action is thus short; they cannot be administered orally as they are destroyed by the digestive process.

Steroid hormones, on the other hand, are usually not stored but are synthesized on demand; these hormones, therefore, do not have a rapid onset of action. Because they have only limited water solubility they can only be transported in the blood, which is predominantly an aqueous medium, if extensively protein bound, and they have a long plasma half-life. Steroids are generally orally active,

**Table 1.1**　A comparison of the properties and actions of peptide and steroid hormones

| Property | Peptide | Steroid |
| --- | --- | --- |
| Synthesis | Stored as inactive precursor from which active hormone is derived | Usually not stored, synthesized on demand |
| Cell membrane permeability | Water soluble, thus unable to cross the cell membrane by diffusion | Highly lipid soluble, thus easily able to cross the cell membrane by diffusion |
| Receptors | Membrane bound receptors on the extracellular surface of the target cell linked either to an ion channel, or to an enzyme capable of generating a second messenger | Intracellular receptors which modulate gene expression and protein synthesis |
| Transport | Transported in solution within the blood | Poorly soluble in aqueous media, thus transported in blood by binding to plasma proteins. Only free, unbound hormone is biologically active |
| Metabolism | Rapidly metabolized by proteolytic enzymes | Slowly degraded by hepatic metabolism, or excreted unchanged. Metabolism may produce other active hormones |
| Plasma half-life | Very short, usually in the order of minutes | Long, usually in the order of hours |
| Duration of effects | Minutes–hours | Hours–days |

although they are extensively metabolized by the liver. Their lipid solubility does however mean that they are able to cross the cell membrane and therefore they have intraceullar receptors, although there is evidence that some steroids are able to produce effects by interaction with membrane bound receptors (see progesterone, Chapter 11). The intracellular receptors produce their effects by influencing protein synthesis, thus the effects of the hormones may take days, or even weeks, to become apparent; however the effects are long lasting, sometimes having a duration of several months.

## Summary

- In order to function correctly the cells of the body must act in a co-ordinated manner; this necessitates intercellular communication. By a process of evolution, two complementary systems of intercellular communication have developed. The first is the nervous system in which cells have developed long projections in order that they may communicate with single, selected, sometimes quite distant cells. The second is the endocrine system which is a much more diffuse network of communication. In this process a hormone is secreted into the blood and dispersed to all cells within the body. The location and number of cells which respond to the hormone is limited by the requirement to possess specific receptors for the hormone in question. Only cells which possess the specific hormone receptors can respond to that hormone.

- Hormones are categorized into two broad chemical groups: the peptides which are not lipid soluble, and the steroids, which are highly lipid soluble. Because of the differences in the properties of the two groups, several different types of hormone receptor have evolved. Peptide hormones interact with membrane bound receptors, and thus they do not need to enter the target cell to exert an effect. These receptors influence the internal processes of the cell by catalyzing the synthesis of an intracellular second messenger, by altering the membrane permeability for specific ions, or by directly altering intracellular enzyme activity. The steroid hormones are able to enter the cell by diffusion across the cell membrane. They are therefore able to produce a direct effect within the cell. This is achieved by influencing gene expression and therefore protein synthesis.

- Disorders of the endocrine system may arise due to excessive or deficient secretion of a hormone, however the receptor population is dynamic. The response of a tissue to a given hormone may decrease by down regulation or desensitization of the hormone receptors; conversely receptors may up regulate. Some endocrine disorders are caused by changes in receptor actvity.

# Selected reading

Brody, T.M., 1994, Sites of action: receptors. In Brody, T.M., Larner, J. and Neu, H.C. (eds), *Human Pharmacology: Molecular to Clinical*, 2nd edn, St Louis: Mosby, 9–23

Jacobs, S.J., 1994, Hormone receptors and signaling mechanisms. In Brody, T.M., Larner, J. and Neu, H.C. (eds), *Human Pharmacology: Molecular to Clinical*, 2nd edn, St Louis: Mosby, 457–471

Mendelson, C.R., 1992, Mechanisms of hormone action. In Griffin, J.E. and Ojede, S.R. (eds), *Textbook of Endocrine Physiology*, 2nd edn, Oxford: Oxford University Press, 28–60

Shenker, A., 1995, G-protein coupled receptor structure and function: the impact of disease-causing mutations. In Thakker, R.V. (ed.), *Baillière's Clinical Endocrinology and Metabolism, International Practice and Research, Vol. 9, No. 3, Genetic and Molecular Biological Aspects of Endocrine Disease*, London: Baillière Tindall, 427–451

# 2 Control of the Endocrine System and the Anterior Pituitary Gland

## 2.1 Introduction

As described in Chapter 1, the endocrine system provides a means whereby long term processes that occur at multiple sites around the body can be controlled simultaneously, for example growth and metabolism. This is very different to the nervous system which provides a very rapid, but very specific means of control. It is important, however, that the actions of the endocrine system and the nervous system remain co-ordinated. Such co-ordination is possible because the part of the endocrine system with greatest influence over all the others is fully integrated with the brain. The hypothalamus is recognized as a neural structure with important roles in the control of sleep and wakefulness, sexual behaviour, hunger and thirst and emotions such as fear, pain and pleasure; but it is also an endocrine gland. It is therefore at the level of the hypothalamus that the activity of the higher centres of the brain are able to influence the activities of the other organs and tissues of the body.

> It is at the level of the hypothalamus that the higher centres of the brain are able to influence the activity of the endocrine system.

## 2.2 Anatomy of the hypothalamus

The hypothalamus lies beneath the thalamus and surrounds the lower portions of the third ventricle (Figure 2.1). It is made up of the floor of the third ventricle which is comprised of the tuber cinereum and the median eminence and several groups of neurones called nuclei, each with different functions, although the precise details of all of their functions are still to be resolved. A continuation of the tuber cinereum and the median eminence forms the pituitary stalk, or infundibulum, which joins the hypothalamus with the pituitary gland (see later).

## 2.3 Secretion of hypothalamic hormones

Many of the nerve cells which make up the hypothalamic nuclei are able to synthesize and secrete hormones. All these cells are under the influence of the other areas of the brain, being controlled in either an excitatory or an inhibitory manner by neurotransmitters such as noradrenaline, 5-hydroxytryptamine, acetylcholine or $\gamma$-amino butyric acid (GABA). All hormones produced by the hypothalamus are peptide in nature, with the exception of dopamine, which is derived from a single

> With the exception of dopamine, which is derived from a single amino acid, all hormones of the hypothalamus are peptide in nature.

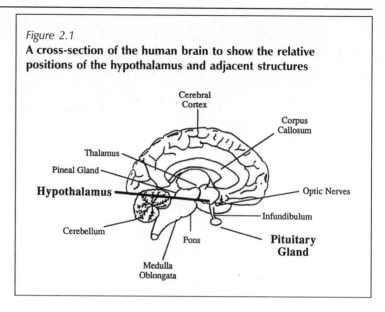

Figure 2.1

**A cross-section of the human brain to show the relative positions of the hypothalamus and adjacent structures**

amino acid. The synthesis of the hormones therefore involves gene expression, protein synthesis by ribosomes and some enzyme activity. Most of the hormones that are synthesized are released by the relevant nerve cells within the area of the median eminence, adjacent to the blood vessels that supply the anterior pituitary gland, the hypothalamicoadenohypophyseal portal system. They are then carried to the anterior pituitary gland where they exert their effect. In the case of the hormones from the supraoptic nucleus and the paraventricular nucleus the hormones are transported directly to the posterior pituiutary gland within the nerve fibres. They are then stored within the gland prior to release.

It is now known that many of the nerve cells within the hypothalamus are able to synthesize and secrete more than one hypothalamic hormone, and that any one single hypothalamic hormone may exert more than one effect within the pituitary gland. This increases the degree of co-ordination in hypothalamic and pituitary gland activity. Further co-ordination of activity is achieved because of the fact that, in many cases, the release of hypothalamic hormones is regulated by some form of negative feedback involving the effects of that hormone (see later). It is also common for the hypothalamic hormones to be released in a pulsatile, rather than continuous manner. These patterns of release reflect the activity of the higher centres of the brain and may be influenced be seasonal or diurnal cycles; the pineal gland and the suprachiasmic nucleus are known to be important in the generation and regulation of many of these secretory rhythms (see later).

The names and functions of the hypothalamic hormones will be discussed when considering the individual hormones of the pituitary gland.

## 2.4 The pituitary gland

The pituitary gland, weighing about 500–800 mg and slightly ovoid in shape, lies beneath the brain within a bony cavity of the skull, the sella turcica. It is connected to the median eminence by the infundibulum or pituitary stalk (Figure 2.2). The gland itself is divided into two lobes which develop from two distinct embryological sources. The anterior lobe is derived from an up-growth of the buccal epithelium, while the posterior lobe is derived from a downgrowth of the brain. The two lobes function generally, but not totally, independently. The anatomy, physiology and pathology of the posterior pituitary gland is discussed in Chapter 3.

The major portion of the pituitary gland is the anterior pituitary or adenohypophysis. This gland is made up of five different types of cell which are named after the hormones that they synthesize. These cells are not distributed uniformly, hence specific regions of the anterior pituitary gland are responsible for the synthesis of specific hormones. The secretory activity of the groups of cells is influenced not only by the hypothalamic hormones, but also by feedback mechanisms involving hormones from the other endocrine glands. Eighty-five per cent of the blood supply for the anterior

*Figure 2.2*

**A detailed enlargement of Figure 2.1 showing how the hypothalamicoadenohypophyseal portal system transports the releasing hormones, which are produced by the hypothalamic neurosecretory cells, from the median eminence to the anterior lobe**

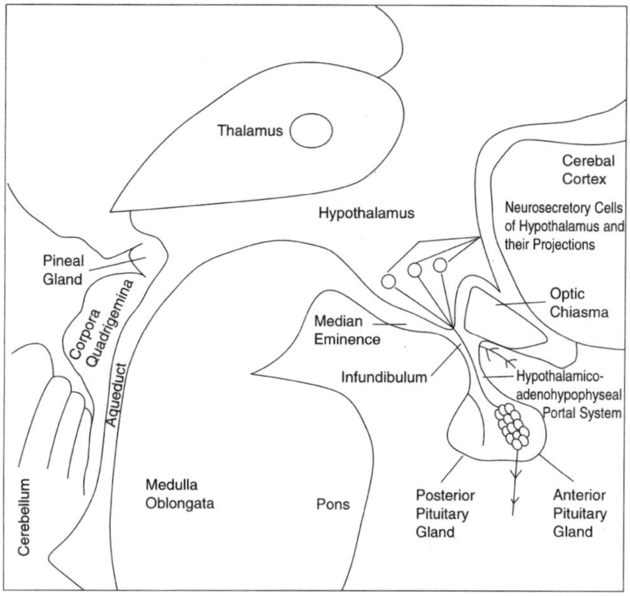

hypothalamico-pituitary is obtained from the hypothalamus via the hypothalamicoadenohypophyseal portal system; the remaining 15 per cent is obtained directly from the superior hypophyseal artery. The blood then drains into the sinus between the meninges.

### 2.4.1  Feedback mechanisms

In order for the endocrine system to function efficiently it must be sensitive to the internal environment and be able to respond to changing conditions. In most cases the endocrine system is controlled by a process called feedback, which may be either negative, or more rarely, positive. Feedback is particularly important in controlling the activity of the hypothalamus and the anterior pituitary gland.

The simplest form of feedback is one in which the secretion of the hormone is directly dependent upon the circulating concentrations of the biochemical, or the activity of the tissue, that is controlled by the hormone in question. An example of this can be seen with the interaction between blood glucose and the hormone insulin (Chapter 7). Insulin causes a decrease in blood glucose concentrations, thus low blood glucose (hypoglycaemia) results in the inhibition of insulin secretion (Figure 2.3). Conversely, high blood glucose (hyperglycaemia) stimulates insulin secretion. In a similar manner, the secretion of hormones of the hypothalamus and the anterior pituitary gland may be regulated by feedback, for example the pituitary hormone thyroid stimulating hormone (TSH) causes the secretion of thyroid hormones by the thyroid gland (Chapter 4). One of the actions of the thyroid hormones is to prevent further secretion of TSH by the anterior pituitary gland. The secretion of TSH is thus controlled by negative feedback. Another example of negative feedback occurs when growth hormone stimulates the secretion of growth hormone release inhibiting hormone (GHRIH, somatostatin) by the hypothalamus, thus decreasing further growth hormone secretion. Some authors cite the example of the control of oxytocin secretion by the posterior pituitary gland (Chapter 3) as a rare example of positive feedback, although in this case the feedback process is very indirect: nipple stimulation due to suckling results in oxytocin secretion; the action of the oxytocin is to cause milk ejection, and therefore continued suckling. The continued suckling results in further oxytocin secretion.

### 2.4.2  Hormones of the anterior pituitary gland

The structure, control of secretion and function of each of the anterior pituitary hormones will be covered separately.

### Thyroid stimulating hormone (TSH)

This glycoprotein hormone, comprised of an $\alpha$ chain of 92 amino acids, non-covalently linked to a $\beta$ chain of 110 amino acids, is

Figure 2.3

**A schematic representation of the different mechanisms of feedback within the endocrine system showing examples of A: direct negative feedback; B: short loop negative feedback and C: long loop negative feedback**

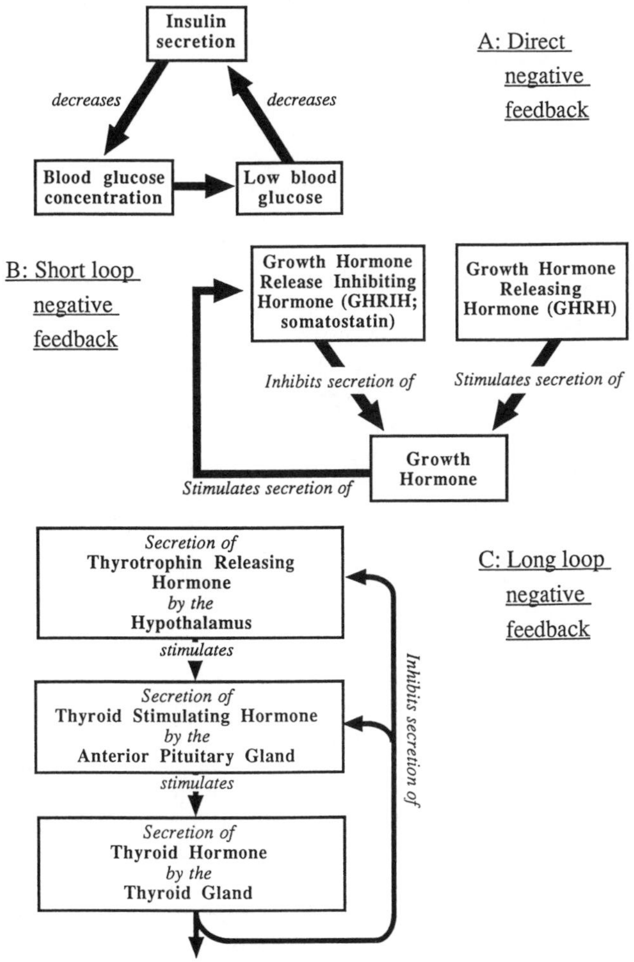

synthesized by the thyrotrophe cells. It is normally secreted in a pulsatile manner with a peak of secretory activity during the night. The principle factor in the stimulation of TSH secretion is the three amino acid hypothalamic hormone thyrotrophin releasing hormone (TRH); oestrogens however also exert an influence as they increase the number of TRH receptors on the thyrotrophes. Secretion of TSH is inhibited by somatostatin and dopamine (see later) and by the negative feedback influence of the thyroid hormones which act both by suppression of TRH secretion and by direct inhibition of TSH secretion.

TSH acts on specific seven transmembrane domain, G-protein linked, receptors on the follicular cells of the thyroid gland to induce synthesis and secretion of the thyroid hormones. This it does by stimulating the uptake of iodine into the follicular cells, stimulating the synthesis of the thyroid hormones and stimulating their secretion. The second messenger for TSH is cAMP.

The physiology of the thyroid hormones and the effects of disorders of TSH and TRH secretion are covered in Chapter 4.

## Adrenocorticotrophic hormone (ACTH)

ACTH is a 39 amino acid peptide hormone which is secreted by the corticotrophe cells. It is secreted in a pulsatile manner with a distinct diurnal rhythm; peak secretion in the early morning, nadir in late evening. The predominant factor in the control of ACTH secretion is corticotrophin releasing factor (CRF), a 41 amino acid hypothalamic hormone. CRF co-exists in some hypothalamic neurones with ADH (see Chapter 3), and this second peptide hormone has also been seen to stimulate ACTH secretion. The stimulus for the release of CRF is stress of any form. ACTH secretion is inhibited by feedback by glucocorticoid hormones. This occurs both at the level of the anterior pituitary gland and in the hypothalamus.

ACTH acts on specific membrane bound receptors in the adrenal cortex to stimulate secretion of the glucocorticoid hormones. The actions of the glucocorticoid hormones and the effects of disorders of ACTH and CRH are discussed in detail in Chapter 9.

## Luteinizing hormone (LH)

LH is one of two glycoprotein hormones which are produced by the gonadotrophe cells of the anterior pituitary. The other hormone is follicle stimulating hormone (FSH), and the two are generally known collectively as the gonadotrophins. LH contains an $\alpha$ chain of 92 amino acids which is identical to that of TSH and FSH, and a unique $\beta$ chain of 115 amino acids. LH does not exhibit diurnal variation in its secretion, but in females its cyclical secretion is central to the control of the menstrual cycle.

The secretion of LH is controlled by the hypothalamic decapeptide gonadotrophin releasing hormone (GnRH) although other messengers such as prolactin and dopamine can influence LH secretion. GnRH is secreted in a pulsatile manner, the pattern of its secretion having an important influence on its effects (see Box 2.1). Oestrogens increase the number of GnRH receptors on the gonadotrophes and therefore enhance its effects.

LH acts on specific G-protein linked receptors of ovarian and testicular target cells to alter the expression of various enzymes. The effects of LH are to stimulate the synthesis and secretion of progesterone and testosterone in females and males respectively, and to induce the process of ovulation. The physiology of the male and female reproductive systems and the role of LH is described in Chapters 10 and 11 respectively.

Box 2.1   **The differential control of the secretion of LH and FSH by the secretion of a single hormone GnRH, by the hypothalamus**

All texts in endocrinology state that the secretion of both FSH and LH by the anterior pituitary gland is controlled by hypothalamic GnRH. In females, however, the secretion of these two pituitary hormones occurs independently, in a cyclical manner (see Chapter 11). The mechanism by which this independent control is achieved is quite complex.

In laboratory animals, a single dose of GnRH causes secretion of LH, but has no effect on FSH; prolonged infusion of GnRH causes secretion of both LH and FSH. This effect occurs because the LH response occurs more rapidly than that of FSH. If there is intermittent, pulsatile infusion of GnRH however, there is an increase in the secretion of FSH, but because the stores of LH become depleted, there is no increase in LH secretion. The hypothalamus can therefore control the secretion of FSH and LH independently by varying the frequency and duration of the pulsatile secretion of GnRH.

The effects of GnRH on LH secretion are inhibited by progesterone and by oestradiol (negative feedback), but six hours after oestradiol treatment the LH secretion is enhanced. Neither progesterone, nor oestrogens when given alone, affect FSH secretion, although if given simultaneously or sequentially – oestrogen followed by progesterone – FSH secretion in inhibited.

During the normal human female menstrual cycle (see Chapter 11) therefore, the early secretion of oestradiol initially inhibits LH secretion, but later enhances it; it is the resultant surge of LH secretion at midcycle that is responsible for ovulation. During the second stage of the menstrual cycle there is secretion of both oestrogens and progesterone thus secretion of LH and FSH is inhibited until the sex hormone secretion subsides and a new menstrual cycle is initiated.

## Follicle stimulating hormone (FSH)

FSH shares the same 92 amino acid $\alpha$ chain as LH and TSH, with a unique $\beta$ chain comprised of 115 amino acids. Like LH, the secretion of FSH is controlled by the hypothalamic hormone GnRH, with the differential secretion of LH and FSH being dependent upon the pattern of the pulsatile GnRH secretion. There is secretion of FSH throughout the female menstrual cycle although the amount secreted varies; secretion of FSH is relatively constant in males. Oestrogens and testosterone enhance the the effects of GnRH on FSH secretion, while inhibin decreases FSH secretion.

FSH acts on specific G-protein linked receptors to stimulate the maturation of ova in females and the subsequent synthesis and secretion of oestrogens. In males FSH acts on the Sertoli cells of the testes to initiate the process of spermatogenesis. The second messenger for the actions of FSH is cAMP. The role of FSH in the physiology of the male and female reproductive systems is decribed in Chapters 10 and 11 respectively.

## Growth hormone (GH, somatotrophin)

Growth hormone exists in several forms, but the major form of GH is a single protein of 191 amino acids; it is structurally similar to prolactin (see later). The secretion of GH is pulsatile and normally shows diurnal variation, the peak secretion occurring during sleep. The stimuli for GH secretion include decreased plasma concentrations of fatty acids and carbohydrates and increases in plasma amino acid concentrations. The pattern of secretion is controlled by two hypothalamic hormones: growth hormone releasing hormone (GHRH), which is a peptide with 40–44 amino acids; and growth hormone release inhibiting hormone (GHRIH), more usually called somatostatin, which is a peptide comprised of 14 amino acids. The secretion of somatostatin is stimulated by activation of noradrenergic and serotonergic pathways within the hypothalamus. GHRH has more effect on GH secretion than somatostatin.

The physiological effect of growth hormone is the initiation and maintenance of linear growth, predominantly by stimulation of amino acid transport and protein synthesis. In some tissues the actions of GH are mediated by the somatomedins. These are proteins of 67 or 70 amino acids which are produced in the liver under the influence of GH. Because of their structural similarity to proinsulin, the somatomedins are now more commonly called insulin-like growth factors I and II. In other tissues growth hormone produces its effects by direct interaction with a dimeric receptor which results in activation of membrane bound phospholipase C or a tyrosine kinase. The physiological effects of growth hormone, its role in the control of growth and development and disorders of growth hormone secretion are described in Chapter 12.

## Prolactin

Prolactin is a peptide, with a structure similar to that of growth hormone, which is comprised of 199 amino acids. The major stimulus for the secretion of prolactin by the lactotrophes of the anterior pituitary gland is the tactile stimulation of the nipple and the surrounding area that occurs during suckling in postpartum, lactating women. Prolactin secretion also occurs in males, although some authors report that it is not initiated by nipple stimulation. Other stimuli that have been shown to cause prolactin secretion include sexual intercourse (cervical stimulation) and stress. Prolactin secretion is so sensitive to the effects of stress that the mild trauma of removing a blood sample from the arm in order to perform the prolactin assay may itself be sufficient to raise prolactin concentrations.

> Prolactin secretion is so sensitive to the effects of stress, that the trauma of giving a blood sample for a prolactin determination may stimulate prolactin secretion.

The secretion of prolactin by the anterior pituitary gland is under the control of the hypothalamus. The predominant controlling factor is dopamine which inhibits prolactin secretion, although secretion is also influenced by TRH which stimulates prolactin secretion. The secretion of prolactin following nipple stimulation thus involves inhibition of dopamine secretion and stimulation of TRH secretion by the hypothalamus. Baseline secretion of prolactin normally occurs in a pulsatile manner and is greatest during sleep. The number of TRH receptors on the lactotrophes is decreased by negative feedback by the thyroid hormones and is increased by oestrogens. Prolactin secretion is thus increased following oestrogen secretion at puberty in females and shows a further rise during pregnancy. Concentrations of prolactin

Figure 2.4

**A schematic representation of the prolactin receptor showing how the prolactin molecule must interact with binding sites on each of the two identical receptor proteins in order to initiate a response**

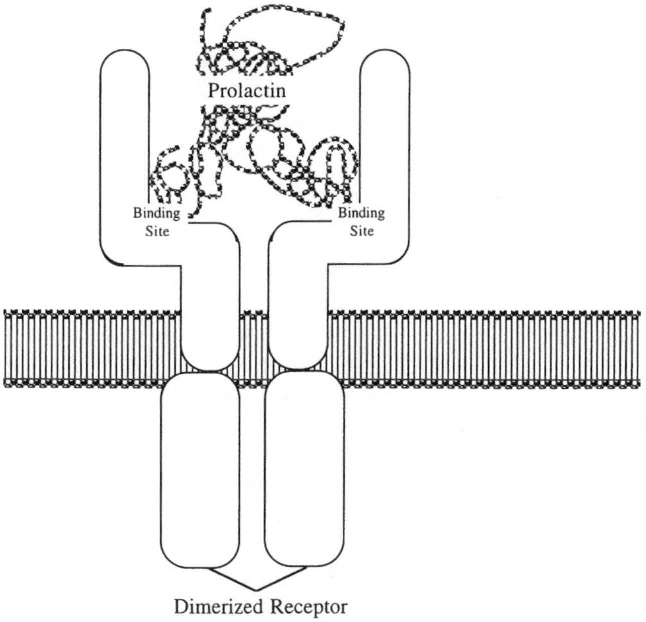

Prolactin

Binding Site

Binding Site

Dimerized Receptor

decline rapidly post-partum if continued nipple stimulation due to breast feeding does not ensue.

The actions of prolactin are mediated by a dimeric tyrosine kinase linked receptor. Two molecules of prolactin are required to activate the two identical receptor molecules which are only active if they occur in pairs (Figure 2.4). The predominant effect of prolactin is to stimulate the growth and development of the breasts and to initiate and maintain milk production. Milk production, however, also requires the presence of hormones such as insulin, corticosteroids and oestrogens. The other effects of prolactin concern the reproductive system, thus prolactin increases the number of LH receptors on the corpus luteum of the ovary and the Leydig cells of the testes. This results in the increased synthesis and secretion of progesterone and testosterone in females and males respectively.

## 2.5 Disorders of the anterior pituitary gland

Disorders of pituitary gland secretion may involve either over- or under-secretion of just one or several of the anterior pituitary hormones. A deficiency of secretion could be caused by a non-secretory tumour of the pituitary, which by expanding compresses the surrounding secretory cells, or it could be due to damage to the pituitary tissue or due to disruption of the vascular links between the hypothalamus and the anterior pituitary. Tumours of the pituitary gland are also often associated with visual disturbances because of compression of the optic nerve by the expanding tumour. It must also be remembered that any damage to or disease of the hypothalamus will be manifested as a disorder of pituitary gland secretion. In the same way, hypersecretion of pituitary hormones may be due to a secretory tumour within the pituitary gland itself, or within the hypothalamus. Tumours of other sites, for example the lungs or the gut, may also produce abnormal amounts of one or more of the anterior pituitary hormones (see Chapter 6).

The symptoms of hypothalamic or anterior pituitary gland disorders will depend on which, and how many of the hormones are secreted abnormally. The symptoms, aetiology and treatment of the conditions caused by abnormal secretion of GnRH and the gonadotrophins, TRH and TSH, CRF and ACTH, and GHRH, somatostatin and growth hormone are covered in the relevant chapters. The following section will discuss only those conditions in which there is abnormal secretion of multiple anterior pituitary hormones and will cover the symptoms associated with abnormalities of prolactin secretion.

Hypersecretion of the anterior pituitary hormones is usually caused by tumours of specific cell types within the pituitary gland itself. The most common form of anterior pituitary tumour is one which secretes prolactin. Commonly if there is hypersecretion of growth hormone it is also accompanied by excess secretion of prolactin. This parallelism is related to the similarities of the two hormones and the similar processes concerned with gene expression and hormone synthesis. Similarly hypersecretion of one of the glycoprotein hormones such as TSH is frequently accompanied by excessive secretion of the other glycoproteins, that is LH and FSH. Each of these conditions would be manifested as a cumulative symptomatology of the individual disorders.

Reduced activity of the pituitary gland is more likely to result in reduced secretion of all of the hormones produced by that gland. Such a condition is called panhypopituitarism. These conditions may result from tumours of the pituitary gland or of the hypothalamus, or may be a result of radiotherapy to treat tumours of the nasopharyngeal area. In some cases the pituitary gland becomes damaged following interruption to its blood supply; this may occur following an aneurysm of the carotid artery or possibly because of the excessive demands of a local tumour depriving the pituitary of its supply of oxygen and nutrients. One special case is Sheehan's syndrome in which panhypopituitarism occurs post-partum. During pregnancy there is an increased blood flow to the pituitary gland because of the tremendous hormonal changes that occur at that time. If, at the time of delivery, there is excessive haemorrhage the blood supply to the pituitary gland may be so disrupted as to cause necrosis of the pituitary tissue. This then becomes manifest as deficiencies of oxytocin, prolactin, antidiuretic hormone and the

thyroid, gonadal and adrenocortical hormones in the weeks following parturition. In panhypopituitarism which presents between the
ages of 20 and 60, the incidence is twice as great in females as it is in
males. In all cases the symptoms of panhypopituitarism are the
accumulation of the symptoms of each of the separate hormones.

### 2.5.1 *Disorders of prolactin secretion*

There is no clinical condition associated with a deficiency of prolactin other than the failure of lactation post-partum. Excess secretion of prolactin – hyperprolactinaemia – however, is
associated with a range of symptoms in both men and women. In
women the most common symptom is amenorrhoea, while in men
the presenting symptom is usually infertility. Between a quarter and
a half of all female sufferers of hyperprolactinaemia also experience
a milky discharge from the breasts. This condition is termed galactorrhoea and it may also occur in up to 30 per cent of male sufferers,
although the most common symptoms in males are decreased
libido, headache and apathy, which are reported by 60–80 per
cent of patients. Of men seeking medical advice for sexual dysfunction, 8 per cent have hyperprolactinaemia. Occasionally hyperprolactinaemia may be associated with gynaecomastia (the
development of feminine breasts in males).

One of the most common symptoms of hyperprolactinaemia is
infertility. Although prolactin is normally involved with the synthesis of LH receptors of the Leydig cells and the corpus luteum, in
excess there is a feedback loop which inhibits secretion of LH by the
anterior pituitary. This in turn leads to menstrual disturbances in
females: amenorrhoea or oligomenorrhoea; and hypogonadism and
infertility in males. This inhibition of fertility by hyperprolactinaemia is the basis of the reduced fertility that occurs while a mother
continues to breast feed her infant. In some societies continued
breast feeding has been used as a deliberate form of family planning.

> Post-mortem studies indicate that 23–27 per cent of the general population have asymptomatic micro-prolactinomas.

There are several causes of hyperprolactinaemia. Prolactin
secreting tumours (prolactinomas) of the anterior pituitary, which
grow slowly over a number of years, are the most common form of
pituitary tumour; post-mortem studies have shown that 23–27 per
cent of the general population have asymptomatic microadenomas
of the lactotrophes at the time of death. The most common cause of
hyperprolactinaemia, however, is probably stress, which can be
either psychological stress or physical stress such as infection.
Deficiency of thyroid hormone secretion may also be associated
with hyperprolactinaemia because of the resultant increase in
TRH secretion as the hypothalamus is freed from negative feedback
and the development of a prolactinoma may occur as part of a
multiple endocrine neoplasia (see Chapter 5). Disruption of dopa-
minergic function is also a common cause of hyperprolactinaemia.
Such disruption follows treatment with dopamine antagonists such
as chlorpromazine and haloperidol which are drugs used in the
treatment of psychotic disorders such as schizophrenia. The anti-

emetic drug metoclopramide is also a dopamine antagonist and has thus been seen to cause hyperprolactinaemia.

The treatment of galactorrhoea, gynaecomastia and hyperprolactinaemia is usually by administration of the dopamine agonist bromocriptine although a new drug, cabergoline, has recently been marketed. These drugs act on the dopamine receptors of the lactotrophes to inhibit further secretion of prolactin; and in approximately 75 per cent of patients prolactin concentrations return to normal and galactorrhoea disappears within eight weeks of starting treatment. With both drugs the most common side effect is nausea, which is a consequence of the stimulation of dopamine receptors of the chemosensory trigger zone of the medulla oblongata. In many cases this treatment is sufficient to limit the growth of any prolactin-secreting tumour, although irradiation of the tissue may be necessary. In cases where the prolactin secretion is a result of drug therapy with a dopamine antagonist the hyperprolactinaemia is resolved by substitution of another form of therapy.

## Summary

- The endocrine system acts in a co-ordinated manner, and complements the actions of the nervous system. This integration of activity is achieved, in part, by the hypothalamus, which is not only an important structure within the brain but is also an important component of the endocrine system. The hypothalamus influences the activity of the endocrine system by the secretion of releasing hormones which are carried within portal blood vessels to the anterior pituitary gland.

- Under the control of the hypothalamus, the anterior pituitary gland secretes many different hormones. The majority of these hormones are termed stimulating, or trophic, hormones, and it is by the secretion of these hormones that the anterior pituitary gland and the hypothalamus are able to control the secretion of thyroid hormones, adrenocorticosteroids and the reproductive hormones. These hormones, in turn, regulate the secretory activity of the hypothalamus and the anterior pituitary gland by a process of negative feedback.

- The anterior pituitary gland also secretes growth hormone and prolactin, which are responsible for the control of growth and milk production respectively. Prolactin also has profound effects on the reproductive system.

- Disorders of the hypothalamus and the anterior pituitary gland result in wide-spread effects throughout the endocrine system which are manifested as symptoms of disorders of each of the associated endocrine glands.

# Selected reading

Ascoli, M. and Segaloff, D.L., 1996, Adenohypophyseal hormones and their hypothalamic releasing factors. In Hardman, J.G., Linbird, L.E., and Gilman, A.G. (eds) *Goodman and Gilman's The Pharmacological Basis of Therapeutics*, 9th edn, New York: McGraw-Hill, 1363-1382

Hadley, M.E., 1992, Pituitary hormones. In Hadley, M.E., *Endocrinology* 3rd edn, Englewood Cliffs: Prentice-Hall, 98–123

Hadley, M.E., 1992, The endocrine hypothalamus. In Hadley, M.E., *Endocrinology*, 3rd edn, Englewood Cliffs: Prentice-Hall, 124–152

Laycock, J. and Wise, P., 1996, Control of the endocrine system. In Laycock, J. and Wise, P., *Essential Endocrinology*, 3rd edn, Oxford: Oxford University Press, 29–39

McCann, S.M. and Ojeda, S.R., 1992, The anterior pituitary and hypothalamus. In Griffin, J.E. and Ojede, S.R. (eds), *Textbook of Endocrine Physiology*, 2nd edn, Oxford: Oxford University Press, 75–105

# 3 The Posterior Pituitary Gland and Pineal Gland

## 3.1 Introduction

In man the posterior pituitary gland secretes two hormones, oxytocin and antidiuretic hormone (ADH). The former is concerned with the release of milk by the breast and with contractions of the uterus, while the latter controls the body's fluid balance by influencing the volume of urine excreted by the kidney. The pineal gland secretes melatonin, which may be involved in the control of circadian rhythms.

## 3.2 Anatomy of the posterior pituitary gland

As described in Chapter 2, the pituitary gland is divided into two anatomically and functionally distinct portions: the anterior pituitary gland (adenohypophysis) and the posterior pituitary gland (neurohypophysis). Embryologically the anterior pituitary gland is derived from an up-growth of buccal ectoderm cells from the roof of the mouth towards the brain while the posterior pituitary gland is derived from neuronal ectoderm cells growing down from the brain. The predominant cells of the posterior pituitary are neuronal cells whose cell bodies lie within the hypothalamus; another cell type, pituicytes, are also present but these are connective tissue cells which are believed to play a mainly supporting role in a manner analogous to that played by glial cells in the brain. The blood supply of the posterior pituitary gland is derived from the inferior hypophyseal artery from whence the blood enters a dense capillary network which is distibuted throughout the posterior lobe of the pituitary gland. The blood eventually drains into the jugular vein, although a small amount may travel to the anterior lobe via a short, interconnecting capillary portal system.

## 3.3 Synthesis and chemistry of posterior pituitary hormones

In man the hormones secreted by the posterior pituitary gland are oxytocin and antidiuretic hormone (ADH). ADH is sometimes known as vasopressin. Both hormones are peptides comprised of nine amino acids, seven of which are identical. In humans ADH has an arginine residue in the penultimate position of the amino terminal and thus this hormone is sometimes called arginine-vasopressin (AVP); in pigs the arginine is substituted by a lysine residue to form lysine-vasopressin (LVP).

Both ADH and oxytocin are produced within the cell bodies of the neurones that make up the posterior pituitary gland. These cell bodies are located within discrete nuclei of the hypothalamus called the supraoptic nucleus (SON) and the paraventricular nucleus (PVN), see Figure 3.1. Each individual neurone synthesizes only one of the hormones, but both oxytocin synthesizing and ADH synthesizing cells are found within each of the nuclei. As with other peptide hormones, oxytocin and ADH are initially synthesized as larger, precursor molecules: pre-pro-oxytocin-neurophysin-I (prepro-OT-NPI) and pre-pro-arginine-vasopressin-neurophysin-II (prepro-AVP-NPII) respectively. Once synthesized, these precursors are stored in granules within the nerve cells. It is within these granules that the precursor hormones are carried, by axonal transport, to the posterior pituitary gland. Each granule contains not only the precursor hormone but also the enzymes necessary for production of the active moiety, this conversion occurs during the two hours taken for the granules to travel from the nerve cell bodies in either the supraoptic nucleus or the paraventricular nucleus, along the hypothalamiconeurohypophyseal

> Both ADH and oxytocin are produced within the supraoptic nuclei and the paraventricular nuclei of the hypothalamus.

Figure 3.1

**The hypothalamiconeurohypophyseal tract transports ADH and oxytocin, which are produced by the nuclei of the hypothalamus, to the posterior pituitary gland**

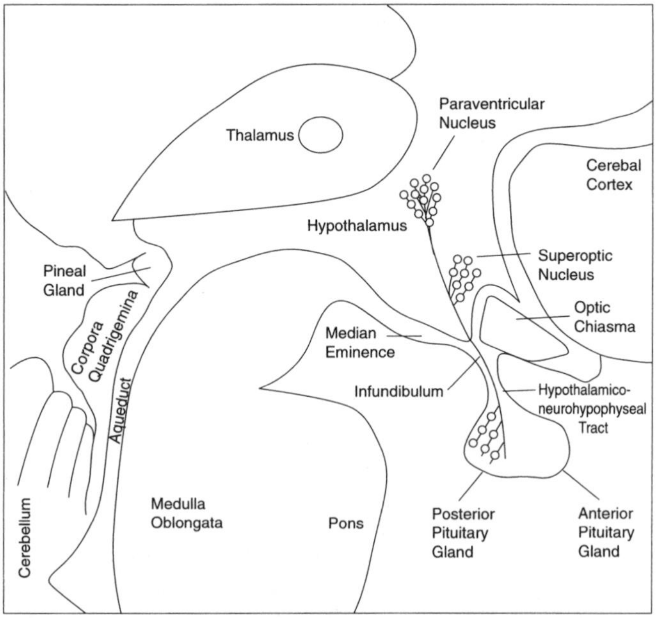

neural tract (which passes through the infundibulum), to the posterior pituitary gland. Within the posterior pituitary gland the ADH or oxytocin is released from the neurone by exocytosis which results in diffusion of the active hormone together with the inactive transport molecule (neurophysin I or II) into an adjacent blood vessel. ADH is also occasionally secreted within the median eminence of the hypothalamus from neurones which are derived from the paraventricular nucleus. The ADH secreted in this region enters the hypothalamicoadenohypophyseal portal system and is transported to the anterior pituitary gland.

## 3.4 Control of posterior pituitary hormone secretion

The stimuli for secretion of oxytocin are nipple stimulation and activation of the stretch receptors of the uterus and vagina.

The release of oxytocin and ADH from the posterior pituitary gland is under independent nervous control. The stimuli for the secretion of oxytocin are nipple stimulation and activation of stretch receptors in the walls of the uterus and vagina. The afferent impulses that are initiated by these stimuli travel along the spinal cord and brain stem eventually to reach the oxytocinergic cell bodies of the supraoptic nucleus and the paraventricular nucleus. Depolarization of the oxytocinergic cell occurs in response to the afferent impulse and an action potential is propagated along the axon. This action potential travels along the hypothalamiconeurohypophyseal tract to the nerve terminal within the posterior pituitary gland, where it results in calcium influx into the nerve terminal which causes extrusion of the hormone containing granules by exocytosis.

Oxytocin secretion is under both excitatory and inhibitory control. At the level of the supraoptic and paraventricular nuclei, afferent impulses may influence the activity of the oxytocinergic cells by the release of either excitatory or inhibitory neurotransmitters. This explains why the secretion of oxytocin can be influenced by emotional state: it is, for example, inhibited during emotional distress. There is also evidence that hormone secretion can be influenced by factors acting within the posterior lobe of the pituitary gland itself, for example enkephalins have be shown to inhibit oxytocin secretion at this point.

A similar situation occurs in the control of ADH release. The major factor influencing the secretion of ADH is plasma osmolarity. Thus an increase in plasma osmolarity stimulates osmoreceptors which are situated within the brain areas around the third ventricle, the circumventricular organs. An important point about the circumventricular organs is that they are not protected by the blood brain barrier, thus they are exposed to the full solute component of the circulating blood. The impulses from the osmoreceptors are carried to the supraoptic and paraventricular nuclei where they cause depolarization of the ADH containing cells. As was seen for oxytocin, the depolarization of these cells ultimately results in calcium

The major factors influencing the secretion of ADH are blood osmolarity and blood volume.

influx into the nerve terminal within the posterior pituitary gland and expulsion of the granules which contain the hormone.

ADH secretion is also influenced by blood volume, thus a secondary mechanism exists whereby stimulation of stretch receptors within the atrium or of baroreceptors within the carotid sinus and aortic arch gives rise to afferent impulses which inhibit ADH secretion. Like oxytocin, secretion of ADH is also influenced by the higher centres of the brain, thus ADH secretion can be altered during times of emotion or stress. The release of ADH within the posterior pituitary gland is also stimulated by endogenous and exogenous opioids and nicotine, and is inhibited following consumption of alcohol.

## 3.5 Mechanisms of posterior pituitary hormone actions

Oxytocin causes smooth muscle contraction by interaction with specific seven transmembrane domain receptors linked to a calcium channel.

Oxytocin causes smooth muscle contraction by interaction with specific seven-transmembrane domain receptors which are linked to calcium channels, thus stimulation of the receptor causes an increase in intracellular calcium concentrations which results in smooth muscle contraction. In some tissues the second messenger for oxytocin is inositol triphosphate but there is also evidence that prostaglandin $E_2$ (PGE$_2$) may be involved: PGE$_2$ enhances the effects of oxytocin on intracellular calcium. The number of oxytocin receptors is increased in the presence of oestrogens and decreased by progesterone, thus oxytocin receptor density fluctuates during pregnancy and during the normal sex cycle (Chapter 11).

ADH also acts on seven-transmembrane domain receptors which are structurally related to the oxytocin receptor. There are two subtypes of ADH receptor, v1 and v2. The v1 receptors are associated with vascular smooth muscle. In these tissues the effect of an interaction of ADH with its receptor is to cause the production of inositol triphosphate and diacyl glycerol. These second messengers in turn cause an increase in intracellular calcium ions leading to smooth muscle contraction. The v2 receptors are found only in the epithelial cells of the renal collecting ducts. Stimulation of these receptors results in the production of the second messenger, cyclic AMP. The second messenger then acts on intracellular enzymes to cause the movement of specific proteins from the cell interior to the cell membrane. The proteins in question are fully formed water channels called aquaporins. Thus the stimulation of the v2 receptor by ADH results in the appearance of water channels in the apical membrane of the renal collecting duct. ADH also causes the production of PGE$_2$ within the kidney, which may be part of a self-regulatory process as PGE$_2$ has been shown to inhibit the effects of ADH on the water channels.

Stimulation of renal receptors for ADH results in the production of cAMP which, in turn, causes the movement of aquaporins into the membranes of the cells of the collecting duct.

# 3.6 Effects of posterior pituitary hormones

## 3.6.1 *Oxytocin*

During pregnancy there is a steady increase in the secretion of oxytocin in response to the normal stimuli. This promotion is a result of the actions of oestrogens on the hypothalamic nuclei; oestrogens also increase the synthesis of oxytocin receptors. Responses to oxytocin are however diminished during pregnancy; this is due partially to the high concentrations of circulating progesterone causing an over-riding inhibition of oxytocin receptor synthesis and, it has been suggested, partially to the increased activity of the enzyme oxytocinase that has been noted at this time. Towards the end of pregnancy there is a decrease in progesterone secretion, thus oxytocin activity becomes more apparent.

The uterine contractions induced by oxytocin are identical to those that occur during normal labour, but oxytocin is not required for a normal labour to progress.

The role of oxytocin in the initiation of normal labour is unclear. There is good evidence that oxytocin secretion rises during labour, and that the uterine contractions induced by oxytocin are identical to those seen during labour, but it has also been reported that in some species the removal of the posterior pituitary gland during pregnancy has no deleterious effect on the process of parturition. It is possible that the uterus remains relaxed during pregnancy because of the presence of the high concentrations of progesterone; however at the end of pregnancy the uterus escapes from the inhibitory influence either because of a slight fall in progesterone secretion during the last two weeks of pregnancy or because of an increase in spontaneous myometrial activity induced by stimulation of uterine stretch receptors caused by the presence of the fully developed foetus (see Figure 3.2). Stimulation of the stretch receptors would also cause secretion of oxytocin, which has the effect of causing further contraction and further stimulation of stretch receptors. The secretion of oxytocin during labour is thus an example of a positive feedback mechanism which continues until the foetus is expelled and the stimulation of the stretch receptors of the uterus and vagina subsides. Prostaglandin $F_2$ also causes uterine contraction and has been shown to increase throughout labour. Thus oxytocin and $PGF_2$ may play complementary roles.

The effects of oxytocin on the non-pregnant uterus may also be important. Vaginal and cervical stimulation, for example during coitus, have been shown to cause oxytocin secretion which results in slow rhythmic contraction of the uterus during the hour after intercourse, these contractions may assist with sperm transport although removal of the posterior gland in animals does not decrease the fertility rate after mating.

In humans the most important effect of oxytocin is the contraction of breast myoepithelium and milk ejection. As described earlier, stimulation of the nipple or the surrounding region during suckling results in secretion of oxytocin (see Figure 3.2). In the lactating breast this results in contraction of the myoepithelial cells surrounding the breast alveoli in which milk is stored. Under normal circumstances the delay between the initiation of

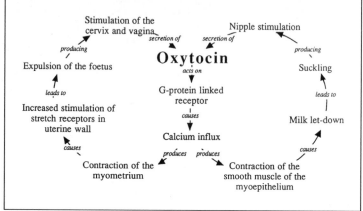

Figure 3.2
The effects of oxytocin on the smooth muscle of the uterus and breast, its mechanism of action and stimulus for secretion. The circular nature of the figure represents the role of positive feedback in the control of oxytocin secretion

suckling, the secretion of oxytocin and the 'let down' or ejection of milk is two to three minutes; however oxytocin secretion can become a conditioned reflex such that it begins in response to the baby's cry or some other stimulus, prior to the initiation of suckling. Again, the oxytocin secretion is under positive feedback control as continued suckling is the stimulus for continued secretion. It is also possible that oxytocin causes enhanced secretion of prolactin (Chapter 2) which would result in production of further milk.

Oxytocin secretion also occurs in males following genital or nipple stimulation, although the function is unknown. It is possible however that effects on reproductive smooth muscle may be involved in ejaculation.

Oxytocin may also play a role in the control of some behaviours. In rats, injection of oxytocin intracerebroventricularly induces full maternal behaviour in animals previously treated with oestrogens while application of oxytocin to the ventromedial nucleus causes female mating behaviour in progesterone pretreated animals.

> Oxytocin secretion can become a conditioned reflex so that it occurs in response to hearing a baby's cry.

### 3.6.2 Antidiuretic hormone

The principle physiological role of ADH is the control of plasma osmolarity. As the sodium concentration of plasma increases, for example due to dehydration, there is a rapid increase in ADH secretion. Changes in blood osmolarity of as little as 1 per cent are sufficient to induce detectable changes in circulating ADH. The released ADH acts on the v2 receptors of the kidney tubule to cause movement of the aquaporin proteins to the apical cell membrane of the collecting duct epithelial cells; this in turn increases the reuptake of water from the tubular fluid and therefore a decrease in urinary volume. ADH can also act on the loop of Henlé to cause increased

Changes in blood osmolarity by as little as 1 per cent, or in blood volume by as little as 8 per cent, are sufficient to stimulate ADH secretion.

sodium reabsorption. At first glance this effect may appear paradoxical, but this action results in an increased concentration of sodium in the tissues surrounding the collecting duct which provides the osmotic gradient for further reabsortion of water (see Figure 3.3).

At normal plasma concentrations ADH has no effect on v1 receptors and thus no effect on blood pressure. If blood volume falls by more than 8 per cent however, for example following haemorrhage, the concomitant drop in blood pressure culminates in reduced stimulation of baroreceptors which results in markedly enhanced ADH secretion. The very high concentrations of circulating ADH that ensue are able to stimulate the v1 receptors and thus cause constriction of the blood vessels of the skin and the splanchnic bed. The overall effect of this vasoconstriction is to increase blood pressure. This action of ADH was the first to be noted, hence the name vasopressin which is still employed by some authors, despite

Figure 3.3

**The effect of increased blood osmolarity or decreased blood volume on the secretion of ADH and the mechanism by which ADH is involved in the maintenance of normal blood volume and pressure**

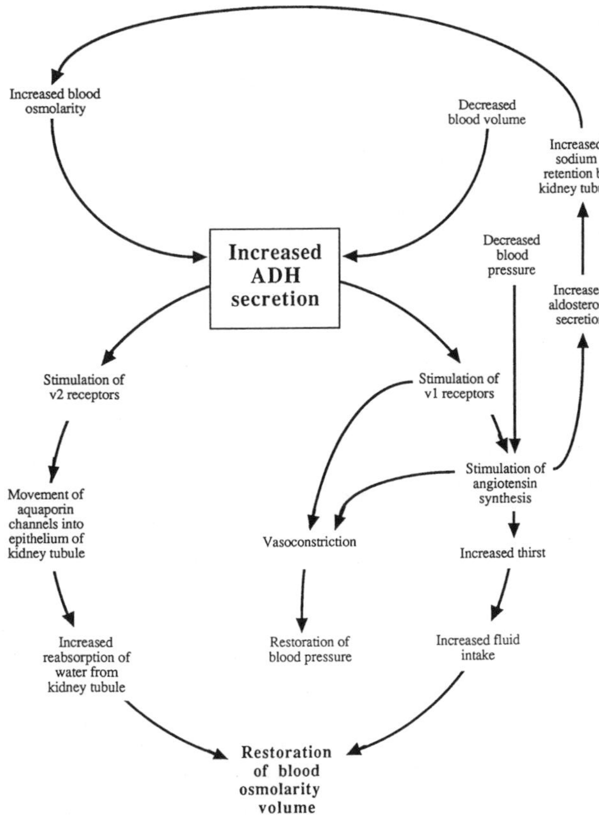

this action being of only limited physiological importance. At high concentrations, ADH may also induce drinking behaviour, possibly by stimulation of the renin–angiotensin system which results in synthesis of angiotensin and the secretion of aldosterone (see Chapter 9). This is another mechanism by which the effects of haemorhage may be mitigated.

ADH, produced by the hypothalamic nuclei but secreted in the medial eminence rather than the posterior pituitary gland, may also have an action in the anterior pituitary gland. It is believed that this ADH enters the hypothalamicoadenohypophyseal portal system which carries it to the anterior pituitary gland. There it is involved in the control of ACTH secretion (see Chapter 2).

## 3.7 Disorders of posterior pituitary hormone activity

### 3.7.1 Oxytocin

There are no recognized clinical conditions associated with either an excess or a deficiency of oxytocin in either males or females, although oxytocin may be used to initiate uterine contractions prior to parturition (see Box 3.1).

### 3.7.2 Antidiuretic hormone

A deficiency of ADH causes diabetes insipidus which is characterized by a failure of the renal collecting duct to reabsorb water. Onset of diabetes insipidus may be very rapid, occurring within two to three days, but the most common cause of this condition is damage to the pituitary gland or hypothalamus due to head injury, although interestingly, even if the pituitary stalk (infundibulum) is severed there is usually sufficient secretion of ADH directly from the cells that remain in the hypothalamus to allow normal control of fluid balance. In a small number of patients there is a familial inability to secrete ADH, usually due to a hypothalamic disorder. In these patients the ADH deficiency is accompanied by diabetes mellitus, optic atrophy and deafness, thus the condition has been labelled DIDMOAD syndrome. Any form of ADH deficiency gives rise to cranial diabetes insipidus; in a related disorder, nephrogenic diabetes insipidus, the kidney fails to respond to the hormone despite normal ADH secretion.

The symptoms of both cranial and nephrogenic diabetes insipidus are excretion of very high volumes of very dilute (pale) urine. The daily volume of urine may reach 20 litres, compared to a normal urinary volume of about 1 litre. There is an accompanying intense thirst. The treatment of cranial diabetes is by administration of the ADH analogue desamino-D-arginine vasopressin (DDAVP). This is usually given either by nasal spray or by injection, although if very high doses are administered orally, sufficient hormone (5 per

> The major symptom of diabetes inspidus is the excretion of high volumes of dilute urine: daily urine excretion may reach 20 litres compared to a normal volume of about 1 litre.

cent) avoids metabolism and is able to produce a therapeutic effect (see Box 3.1).

Administration of ADH or its analogue is useless in the treatment of nephrogenic diabetes insipidus. In this condition it is possible to reduce the excessive urine excretion by administration of thiazide diuretics which are believed to exert their effect by reducing glomerular filtration rate.

Ectopic secretion of ADH may occur in patients with a variety of tumours including tumours of the lungs, colon, prostate and adrenal cortex. This phenomenon may reflect the common embryological origin of the hypothalamic cells which produce the ADH and the cells of the tumours. Excess ADH secretion may also occur following head injury or accompanying certain infections. The condition is usually termed syndrome of inappropriate ADH (SIADH). The symptoms of the condition are fluid retention coupled with decreased plasma sodium concentrations. Treatment is by limitation of fluid intake (500 ml per day), although there have been trials with drugs such as the antibiotic demeclocycline which prevent the actions of ADH at its receptor and therefore induce temporary 'nephrogenic diabetes insipidus'.

## 3.8 The pineal gland

The pineal gland is a small gland which is attached to the roof of the third ventricle and lies between the cerebral cortices and the cerebellum. It is the only unpaired structure of the brain which led Descartes to suggest that it was the seat of the soul. The pineal gland receives a very rich blood supply, and lies outside the blood–brain barrier and its activity is controlled by noradrenergic innervation arising from the suprachiasmic nuclei; its activity is

*Box 3.1* **Currently available preparations of posterior pituitary gland hormones and drugs for the treatment of disorders of the posterior pituitary gland**

|  | *Drug* | *Proprietary name* |
|---|---|---|
| Drugs used in the treatment of diabetes insipidus | Vasopressin | Pitressin (injection) |
|  | Desmopressin | Desmotabs (tablets) |
|  | Desmopressin | Desmospray (nasal spray) |
|  | Lypressin | Syntopressin (nasal spray) |
|  | Terlipressin | Glypressin (injection) |
| Drugs used in the treatment of syndrome of inappropriate ADH secretion | Demeclocycline | Ledermycin |
| Drugs used for the induction of labour | Oxytocin | Syntocin (injection) |

thus influenced by the response of the retina to light. In some species the pineal gland itself shows a direct response to light, which has led to it being labelled as the 'third eye', although in humans this is not the case. The principal secretion of the pineal gland is melatonin, a hormone derived from 5-hydroxytryptamine (serotonin), although there is also limited secretion of thyrotrophin releasing hormone and thyroid hormones. These hormones are secreted rhythmically, with greatest secretion occurring during times of darkness. The actions of melatonin have been associated with various forms of circadian activity, thus it may be involved in the seasonal and diurnal variations of skin and coat pigmentation that occur in some species. It may also be responsible for the fluctuations of reproductive activity that occur in many species, with melatonin having an antigonadotrophic effect. In humans it has been suggested that melatonin may be involved in the timing of puberty, the suggestion arising from the observation that melatonin secretion declines after puberty, and that some cases of precocious puberty are associated with pinealomas. Melatonin has also been implicated in the aetiology of the symptoms of jet lag, and seasonal affective disorders (SAD), but as yet there have been no therapeutic ramifications of these suggestions.

## Summary

- The posterior pituitary gland is derived from a down-growth of the neural tissue of the brain and acts independently of the anterior pituitary gland. Unlike other endocrine glands, the posterior pituitary gland does not synthesize hormones, but rather stores hormones synthesized within the hypothalamus. The hormones released from the posterior pituitary gland, antidiuretic hormone (ADH) and oxytocin, are both peptide hormones, with highly related structures, which act on similar G-protein linked receptors.

- Oxytocin acts to cause uterine contractions and milk ejection in females; its function in males is unknown. Synthesis of oxytocin increases during pregnancy, however its actions are inhibited by a decrease in oxytocin receptor density and an increase in the activity of oxytocinase. Oxytocin is secreted following stimulation of the cervix and vagina or of stretch receptors of the uterine wall. In late pregnancy the uterine contractions induced by oxytocin are indistiguishable from those of normal labour, however oxytocin is not essential for parturition. Oxytocin secretion may also be induced by stimulation of the nipples in both males and females; this is the process by which milk ejection is induced in lactating females.

- ADH stimulates reabsorption of water by the collecting duct of the kidney. The stimulus for ADH secretion is an increase in plasma osmolarity. Following haemorrhage ADH also causes vasoconstriction, which results in restoration of normal blood pressure. A deficiency of ADH function results in diabetes insipidus which is characterized by excess excretion of very dilute urine and an accompanying thirst. Excess or ectopic ADH secretion results in excessive fluid retention.

- The pineal gland is situated between the cerebellum and the cerebral cortices. The principal secretion of the pineal gland is melatonin which is concerned with circadian rhythms.

## Selected reading

Binkley, S.A., 1995, Vasopressin and oxytocin. In Binkley, S.A., *Endocrinology*, New York: HarperCollins Publishers, 88–106

Hadley, M.E., 1992, Neurohypophyseal hormones. In Hadley, M.E., *Endocrinology*, 3rd edn, Englewood Cliffs: Prentice-Hall, 153–178

Samson, W.K., 1992, The posterior pituitary and water metabolism. In Griffin, J.E. and Ojede, S.R. (eds), *Textbook of Endocrine Physiology,* 2nd edn, Oxford: Oxford University Press, 106–117

# 4 The Thyroid Gland

## 4.1 Introduction

The thyroid gland, and its secretions, are not essential for life, but they are essential for normal development and for the physical and mental well-being of the individual. The hormones secreted from this gland are the thyroid hormones and calcitonin, which function independently and which are under independent control. This chapter will consider the physiology and pathology of the thyroid hormones; calcitonin and its role in the control of calcium homoeostasis is discussed in Chapter 5.

## 4.2 Anatomy of the thyroid gland

In total, the diffuse thyroid gland weighs about 25 g and has the size and shape of a bow-tie. It is situated in the neck and lies over and around the trachea; it is one of the few endocrine glands that lie so superficially that changes in size are manifested in changes in surface anatomy, for example an enlargement of the thyroid gland produces a marked swelling of the neck.

In the embryo the thyroid gland is formed from the pharynx, but by the second month of gestation it is anatomically distinct from the alimentary tract. Ultimately the gland develops into a highly vascular, bilobular structure, with one lobe lying either side of the trachea and the two lobes being connected by a narrow isthmus. Each lobe of the gland receives a supply of blood from the external carotids via the superior thyroid arteries and from the subclavian arteries via the interior thyroid arteries. At the microscopic level, the thyroid gland is composed of follicles. Each follicle consists of a monolayer of epithelial cells enclosing a large core of viscous, homogeneous colloid. A single follicle may have a diameter of up to 1 cm. During times of increased thyroid gland activity there may be a decrease in the volume of colloid within each follicle, while glandular inactivity is accompanied by increased glandular volume. This indicates that the colloid acts as a reservoir of thyroid hormone. Within the walls of the follicles, and occasionally within the extra-follicular spaces, there are a small number of clear cells, called paraventricular or C cells; these cells produce the other hormone of the thyroid gland, calcitonin.

# 4.3 Synthesis and chemistry of thyroid hormones

Thyroid hormones are derived from the amino acid tyrosine by a process which involves the incorporation of iodine. The synthesis of these hormones begins in the foetus at about the third month of gestation and continues fairly constantly throughout life although there is a surge of thyroid gland activity at puberty and during pregnancy.

The first stage in the synthesis of the thyroid hormones is the uptake of iodide by the follicular cells. Iodide is obtained from the diet and is taken into the follicular cells by an active pump mechanism, similar to the $Na^+K^+$ ATPase sodium pump, which transports iodide against an electrochemical gradient. By the activities of this pump, the iodide concentration within the follicular cells is approximately 25–50 times greater than the concentration in plasma. Once in the cell, the iodide is oxidized by peroxidase after which it becomes covalently bound to tyrosine residues within thyroglobulin. Thyroglobulin is a glycoprotein, of molecular weight 660 000, which is synthesized within the follicular cells. Initially the iodination of tyrosine results in the formation of either monoiodotyrosine (MIT) or diiodotyrosine (DIT); however a further coupling reaction occurs which results in the combination of two DIT molecules to form tetraiodothyronine ($T_4$, or thyroxine) or the combination of MIT with DIT to form triiodothyronine ($T_3$) (see Figure 4.1). These iodothyronines, still associated with thyroglobulin, are then secreted into the lumen of the follicle to form the colloid.

Prior to secretion of the thyroid hormones, colloid is moved from the lumen of the follicle into the follicular cells by endocytosis. The

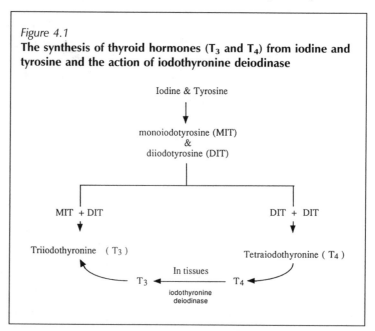

Figure 4.1

**The synthesis of thyroid hormones ($T_3$ and $T_4$) from iodine and tyrosine and the action of iodothyronine deiodinase**

Iodine & Tyrosine

monoiodotyrosine (MIT)
&
diiodotyrosine (DIT)

MIT + DIT

DIT + DIT

Triiodothyronine  ($T_3$)

Tetraiodothyronine ($T_4$)

In tissues

$T_3$ ← $T_4$

iodothyronine deiodinase

Approximately 95 per cent of the thyroid hormone leaving the thyroid gland is in the form of tetraiodothyronine.

colloid resides within the cytoplasm of these cells in the form of droplets or granules. Once in the cytoplasm, the action of proteases on the colloid results in the degradation of the thyroglobulin and the liberation of the iodinated tyrosine moieties. MIT and DIT are rapidly degraded by halogenases to free the iodide, which is then re-utilized by combination with thyroglobulin. The $T_3$ and $T_4$ leave the follicular cells and enter the bloodstream for distribution to the target tissues. Approximately 95 per cent of the thyroid hormone leaving the thyroid gland is in the form of $T_4$ (thyroxine; the normal rate of secretion of this hormone is approximately 100 nmol/day.

Thyroid hormones are insoluble in water, therefore in order for them to be transported within the blood they must be bound to plasma proteins. Over 99 per cent of the circulating thyroid hormones are protein bound, the majority (approximately 75 per cent) of $T_4$ binding is to thyronine binding globulin (TBG), with a further 15–20 per cent being bound to thyroxine binding prealbumin (TBPA) and 5–10 per cent being bound to albumin. In the case of $T_3$ there is extensive binding to TBG and albumin, with very little binding to TBPA. The total plasma protein binding of $T_4$ is greater than that of $T_3$, thus $T_3$ accounts for approximately 20 per cent of the free thyroid hormones. It should be remembered that only free, unbound hormone is biologically active, thus any factor which alters the extent of protein binding of these hormones can have profound effects on their apparent biological activity.

Because of their lipid solubility, thyroid hormones are able to cross the cell membranes of the target cells easily, although some target cells possess an active transport mechanism for the uptake of both $T_3$ and $T_4$. Within the target tissues deiodinase enzymes convert the $T_4$ to either $T_3$ (80 per cent) or reverse-$T_3$ (20 per cent). $T_3$ has a biological activity approximately 40 times greater than that of $T_4$, while reverse-$T_3$ is biologically inactive. Some of the $T_3$ which is produced re-enters the bloodstream, while the majority acts within that tissue; up to 90 per cent of the biologically active thyroid hormone within the cell is in the form of $T_3$. These facts may suggest that the function of $T_4$ is merely as a precursor for the production of $T_3$. There is however evidence firstly, that $T_4$ has its own specific receptor, and secondly, that cells can moderate the proportion of $T_3$ and reverse-$T_3$ produced from $T_4$ in order to control the local responses to the circulating thyroid hormones. The plasma half-life of $T_4$ is six to eight days while that of $T_3$ is one day.

At the cellular level, up to 90 per cent of the biologically active thyroid hormone is in the form of triiodothyronine.

## 4.4 Control of thyroid hormone secretion

The synthesis and the secretion of the thyroid hormones is under the control of thyroid stimulating hormone (TSH, or thyrotrophin) which is secreted from the anterior pituitary gland. The release of TSH is stimulated by thyrotrophin releasing hormone (TRH), which is a tripeptide hormone synthesized in the paraventricular nuclei of the hypothalamus. Somatostatin and feedback by the thyroid hormones reduce the effects of TSH (see Figure 4.2). TSH is a

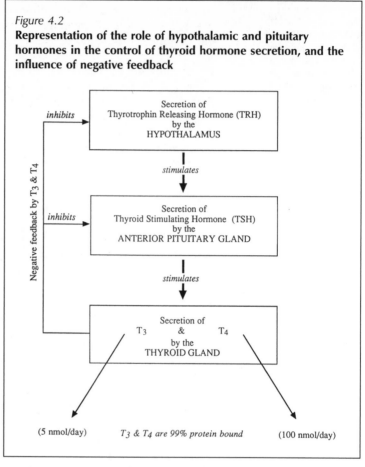

*Figure 4.2*

**Representation of the role of hypothalamic and pituitary hormones in the control of thyroid hormone secretion, and the influence of negative feedback**

glycoprotein hormone which produces its effects by interaction with membrane bound receptors on thyroid follicular cells and the subsequent production of the second messenger cAMP. TSH stimulates all of the stages of thyroid hormone synthesis by increasing the production of thyroglobulin, by stimulating the uptake of iodide into the follicular cell both by increasing the number of iodide pumps and by increasing their activity, and by increasing the incorporation of iodide with the tyrosine moities of the thyroglobulin. TSH also causes thyroid hormone secretion by increasing the rate at which colloid is taken back up into the cytoplasm and the rate at which the $T_3$ and $T_4$ is liberated. In the absence of TSH the thyroid gland atrophies, while prolonged TSH secretion results in thyroid hyperplasia. Under normal circumstances the release of TSH is pulsatile with peaks at night; the plasma half-life of TSH is about one hour.

## 4.5 Mechanism of thyroid hormone action

The receptors for the thyroid hormones are intracellular, more specifically nuclear. Thus the hormones must first cross the cell membrane before they can produce their effect. The effect of the inter-action of the thyroid hormones with their receptors is to influence gene transcription and thus protein synthesis; increased synthesis of several enzymes, and even $Na^+K^+$ ATPase, has been demon-strated following stimulation of the thyroid hormone receptors. To date, two separate thyroid hormone receptors have been identified although specific functions of the receptor subtypes have yet to be elucidated. Both receptors have a much greater affinity for $T_3$ than for $T_4$.

In addition to the genetic effects, it has also been suggested that thyroid hormones may be able to produce effects by direct actions on mitochondria and membrane transport proteins.

## 4.6 Effects of thyroid hormones

Qualitatively, there are no differences in the effects of $T_3$ and $T_4$. The main effect of the thyroid hormones is to increase basal meta-bolic rate; this involves an increase in carbohydrate metabolism and an increase in the synthesis, mobilization and degradation of lipids. Protein synthesis is also stimulated, hence thyroid hormone is essential for normal growth. Thyroid hormones are also essential for the normal development of the CNS, especially myelination of nerve fibres.

**The main effect of the thyroid hormones is to increase basal metabolic rate.**

Thyroid hormones increase basal metabolic rate, and therefore oxygen consumption, in nearly every organ, the major exceptions being the brain, uterus, testes and spleen. Interestingly the thyroid gland and the anterior pituitary gland are also refractory to the effects of thyroid hormones on metabolism. The primary mechan-ism behind this effect is an increase in the number and size of mitochondria and an increased activity of metabolically important enzymes. All aspects of carbohydrate metabolism are increased so that there are increases in glycogenesis and glucose uptake by muscle cells and adipose cells, possibly by a potentiation of the effects of insulin, but also increases in glycogenolysis, gluconeo-genesis and glycolysis, the latter possibly due to a potentiation of the effects of catecholamines. There is also an increase in glucose absorption by the gastrointestinal tract. The potentiation of the effects of catecholamines may also explain the increased rate of lipolysis, which predominates over the effects on lipogenesis, the rate of which is also increased (see Box 4.1).

The thyroid hormones also have effects on protein and bone. The synthesis of many proteins is dependent upon the presence of thyroid hormones, hence thyroid hormone is essential for normal growth and development. In some animals these hormones control metamorphosis. A deficiency of the thyroid hormones may result in an impairment of normal growth, however an excess has similar effects due to the stimulation of gluconeogenesis promoting protein catabolism. A similar picture is seen in bone where thyroid hor-mones are involved in both bone synthesis and resorption; thus an excess secretion promotes not only bone resorption and

Box 4.1   **The actions and effects of thyroid hormones**

| Mechanism of action | Effects of thyroid hormones |
|---|---|
| Thyroid hormones act on intracellular receptors to influence RNA activity; this affects:<br><br>  protein synthesis<br>  enzyme synthesis<br>  oxidative phosphorylation | The main effect of thyroid hormone action is to increase metabolic rate and oxygen consumption in tissues throughout the body.<br><br>Thyroid hormones are also involved in growth and maturation. |

demineralization but also a breakdown of the protein matrix of the bone.

The potentiation of the effects of catecholamines by thyroid hormones is mentioned above. These effects may be explained by an increased expression of $\beta$-adrenoceptors by adipocytes. As well as the effects of this action on carbohydrate metabolism, there are also effects on the cardiovascular system. Thyroid hormones increase cardiac output and sometimes increase the force of contraction of the heart and also increase the rate of blood flow through surface vessels. It is probable, however, that these effects are not solely due to increases in the sensitivity of the $\beta$-adrenoceptors but are partially brought about by the increased metabolic demands of the tissues and by a direct action of the thyroid hormones on cardiac muscle.

An action of the thyroid hormones on catecholamine receptors may also explain some of the effects of these hormones in the central nervous system. Thyroid hormones affect mental activities such as arousal and memory in adults and in children there are effects on intellectual ability. The latter effects may, in part, be a consequence of the requirement of the thyroid hormones for the normal development of cortical neurones and the myelination of nerve fibres.

# 4.7 Effects of thyroid dysfunction

## 4.7.1  *Hypoactivity*

The symptoms of hypothyroidism are of slow onset and are generally those of decreased basal metabolic rate. The classical features of the disorder are decreased heart rate with decreased respiration rate and body temperature. There may be some weight gain due to the decreased utilization of carbohydrate and fat stores, and reflexes are slowed due to decreased neuronal function. Further symptoms of the decreased neuronal function are the feelings of tiredness and lethargy and possible psychological disorders such as depression. There may also be dry, flaky skin with hair loss and oedema. The oedema, which is typically widespread, is due to accumulation of skin proteins, polysaccharides and hyaluronic acid in the subcutaneous spaces. It is this oedema which resulted in the condition being named myxoedema. Another feature of this condition is a deepening of the voice, possibly as a consequence of the oedema around the larynx. Because of disruption of other endocrine systems, hypothyroidism also causes reproductive disorders, with menorrhagia or irregular menstruation in females and reduced fertility in both males and females (see Box 4.2).

If decreased thyroid activity occurs *in utero*, or in the neonate, there is the risk of severe mental retardation due to the failure of development of the central nervous system. This is the condition known as cretinism; other features include short stature and oedema, coupled with a protruding tongue, which give rise to coarse features and a hoarse cry. Cretinism occurs in about 1:4000 births and is usually due to a failure in the development of the thyroid gland. It is irreversible if untreated within a few weeks of birth, however treatment is so successful if the condition is recognized rapidly that many countries routinely screen for the condition by assays of TSH in the newborn.

In addition to the features of thyroid hypoactivity that are seen in adults, if the condition occurs in children there are also deficiencies in growth and maturation. Typically there is delayed sexual maturation coupled with diminished linear growth and reduced limb length to trunk length ratio (see Chapter 12).

Symptoms of hypoactivity of the thyroid gland may be a consequence of an inability of the thyroid gland to synthesize the thyroid hormones due to a deficiency of iodide or possibly due to a lack of the enzymes required for hormone synthesis. In some cases it may be a dysfunction of secretion of TSH by the anterior pituitary gland or of TRH by the hypothalamus that is the cause of the problem. Alternatively the target tissues may fail to respond to thyroid

*Box 4.2* **The features of hypothyroidism**

| | |
|---|---|
| Prevalence: | 0.01–0.08 per cent of population; 80 per cent of sufferers are female, commonly 30–60 years of age |
| Causes: | Autoimmune disease (e.g. Hashimoto's disease)<br>Gland destruction<br>Cancers<br>Some drugs |
| Symptoms: | Dry skin<br>Hair loss<br>Retarded growth<br>Muscle stiffness<br>Lethargy, bradycardia<br>Constipation<br>Goitre (see below) |
| Treatment: | Thyroxine replacement |

A **goitre** is a swelling of the neck. It can occur in hypothyroidism, if there is a compensatory increase in TSH secretion, or in hyperthyroidism if there is excessive secretion of TSH hormones due to a congenital defect in the expression of the thyroid hormone receptors. The most common cause of hypothyroidism, however, is Hashimoto's disease. This is an autoimmune disorder in which there is cell mediated immune response resulting in the destruction of the follicular cells of the thyroid gland, or occasionally an immune response resulting in the production of an antibody capable of blocking the TSH receptor; circulating antithyroid antibodies are detected in 80 per cent of cases of hypothyroidism. Hashimoto's disease affects approximately 1 per cent of the population and it is ten times more common in females than in males.

Hashimoto's disease affects approximately 1 per cent of the population.

The treatment of thyroid hypoactivity is relatively straightforward if the condition is due to hormonal hyposecretion. In the case of iodine deficiencies, supplemention of iodine is effective, although administration of excessive doses of iodine paradoxically results in further decreases in thyroid hormone synthesis (see later). The most common form of treatment is lifelong hormone replacement therapy with oral $T_4$. $T_4$ is favoured over $T_3$, firstly on the grounds of cost, and secondly because its longer plasma half-life means that stable plasma concentrations can be achieved after once daily dosing (see Box 4.3). In most cases the dose must be titrated against response and is therefore subject to constant monitoring. Administration of excessive replacement doses may cause cardiovascular problems and other symptoms of hyperthyroidism.

At present there are no known treatments for hypothyroidism due to thyroid hormone receptor resistance.

### 4.7.2 *Hyperactivity*

The symptoms of hyperthyroidism usually develop slowly and are those of increased metabolism. There is tachycardia and possible cardiac arrhythmias due partly to the increased metabolic demands of the tissue and partly due to the increased number of cardiac $\beta$-adrenoceptors; excessive sweating due to a combination of the increased body temperature and the increased activity of the sympathetic nervous system; agitation and anxiety, again possibly related to the increased $\beta$-adrenergic activity and weight loss despite increased appetite due to the catabolism of carbohy-

Box 4.3   **Drugs used for the control of thyroid disorders**

| Thyroid hormones are used in replacement therapy of hypothyroidism: | | Antithyroid drugs are used in the treatment of hyperthyroidism and for the preparation of patients for thyroidectomy: | |
|---|---|---|---|
| *Drug* | *Proprietary name* | *Drug* | *Proprietary name* |
| Thyroxine | Eltroxin (tablets) | Carbimazole | Neo-Mercazole |
| Triiodothyronine | Tertroxin (tablets) | | (tablets) |

drates, lipids and proteins. Osteoporosis and amenorrhoea may also occur and in nearly all cases there is a goitre (see later).

The most common cause of hyperthyroidism is Graves' disease which affects approximately 1 per cent of the adult population (2 per cent of adult females and 0.2 per cent of adult males). Graves' disease is an autoimmune disease in which there is an antibody to TSH receptors on the thyroid gland. The action of this antibody is to stimulate the receptor and therefore to potentiate the synthesis and secretion of thyroid hormones; the antibody responsible for this effect has been labelled variously as the long-acting thyroid stimulating (LATS) antibody, thyroid stimulating immunoglobulin (TSI) and thyroid stimulating antibody (TSAB), although whether these names refer to the same immunoglobulin is unclear. Other antibodies produced as a consequence of Graves' disease are responsible for the other features of the disorder, most notably the exophthalmos or proptosis induced by retro-orbital oedema and fat deposition and the pretibial myxoedema due to subdermal mucinous deposits. It therefore follows that the ophthalmic and myxoedematmous symptoms of Graves' disease are due to the autoimmune component of the disorder rather than the hyperthyroidism. Together with toxic nodular goitre, which is overactivity of a single 'nodule' of the thyroid gland, Grave's disease comprises 99 per cent of cases of hyperthyroidism (see Box 4.4).

> Graves' disease affects 2 per cent of adult females and 0.2 per cent of adult males.

Other causes of hyperthyroidism include TSH producing tumours of the anterior pituitary gland, TRH secreting tumours of the hypothalamus, Jod–Basedow phenomenon, in which iodine administration results in excessive secretion of the thyroid hormones, hyperthyroidism of pregnancy and the puerperium, and a condition in which the TSH receptors are constituitively active. In

Box 4.4 **The features of hyperthyroidism**

| | |
|---|---|
| Prevalence: | 0.2–0.4 per cent of population, ratio of women : men is 8:1 |
| Causes: | Graves' disease (most common)<br>Iodine overdose<br>Thyroid hormone overdose<br>Thyroid carcinoma<br>TSH secreting tumour (accompanied by acromegaly)<br>TRH secreting tumour |
| Symptoms: | Increased body temperature<br>Raised blood pressure<br>Weight loss in spite of raised appetite<br>Hyperactivity |
| Treatment: | Antithyroid drugs<br>Surgery<br>Radioactive iodine ($^{131}$I) |

Note: Exophthalmos is a symptom of Graves' disease *not* hyperthyroidism

the latter condition there is a genetic defect in the receptors such that they generate the second messenger in the absence of the stimulating hormone. Hyperthyroidism may also occur as part of a multiple endocrine neoplasia (see Chapter 5) or following ectopic secretion of thyroid hormones from an ovarian tumour (teratoma). Thyroid cancers are the most common endocrine tumour (five per million of the population), but in England and Wales they only account for 300 deaths per year.

The treatment of Graves' disease and other forms of hyperthyroidism is aimed at the reduction of circulating thyroid hormones although many of the symptoms of hyperthyroidism can be relieved by use of $\beta$-adrenoceptor antagonists. The treatment options are surgical removal or reduction of the thyroid gland, destruction of thyroid tissue using radioactive iodine or reduction of the synthesis of thyroid hormones using antithyroid drugs (see Box 4.3). The other symptoms such as exophthalmos are sometimes treated by the use of immunosuppressant doses of corticosteroids (see Chapter 9).

The antithyroid drugs carbimazole and methimazole act by inhibition of the synthesis of thyroid hormones by preventing the incorporation of the iodide into the thyroglobulin; in addition to this effect propylthiouracil prevents the peripheral conversion of $T_4$ to $T_3$. Potassium perchlorate acts by competing with iodide for the active iodide uptake pump. Drug therapy is usually continued for 12 to 24 months; careful selection of the dose allows normal thyroid hormone concentrations to be achieved but remission occurs in about half of the patients when drug therapy is stopped. Preparations currently available in the UK are Peroidin (potassium perchlorate) and Neo-Mercazole (carbimazole).

Surgical removal of part of the thyroid gland is only considered in cases where treatment with antithyroid drugs has failed or where there is a large goitre (see later). Complications can occur due to the removal of the parathyroid glands (see Chapter 5) and in up to 20 per cent of cases the hyperthyroidism persists after surgery. Excessive removal of tissue leads to hypothyroidism in up to 60 per cent of cases. An alternative to surgery is the use of radioactive iodine ($^{131}$I). This is selectively concentrated in the thyroid gland where it causes tissue damage and therefore reduced thyroid hormone secretion. This treatment is not commonly used in young people or women of child bearing age due to the fears of teratogenic and carcinogenic effects, although the treatment has been in use for over 40 years without adverse effects being apparent.

Goitre, an enlargement of the thyroid gland, may be associated with either hypoactivity or hyperactivity of the thyroid gland.

### 4.7.3  *Goitre*

A goitre is an enlargement of the thyroid gland to produce a swelling of the neck. Goitre may be a symptom of either an overactive or underactive thyroid gland. In the case of underactivity (e.g. due to iodine deficiency), the anterior pituitary gland will secrete increased quantities of TSH in an attempt to boost thyroid hormone synthesis and secretion. The thyroid gland enlarges under the influence of the TSH. In the case of overactivity of the thyroid gland, the error may lie in the hypothalamus or anterior pituitary gland. Excessive production of either TRH or TSH will lead to an enlargement of the thyroid gland (goitre) accompanied by excessive thyroid hormone secretion. A goitre may also be caused by a tumour or infection of the thyroid gland. In some areas, for example New Guinea, the Andes and the Himalayas, there is a high incidence of endemic goitre due to a paucity of iodine in the diet. Prophylactic administration of iodine either by injection or by incorporation into table salt or flour has markedly reduced the incidence of endemic goitre worldwide, although it carries with it the risk of Jod–Basedow phenomenon in which iodine administration precipitates hyperthyroidism. There are also drugs which can induce goitre, most notably lithium which is used in the treatment of manic depression, and certain iodides which are contained in vitamin preparations and some cough remedies. These ions are selectively concentrated within the thyroid gland where they interfere with iodide incorporation and hormone release. Goitre is also common at puberty and during pregnancy due to the excessive hormonal flux at that time.

Many goitres are self limiting or respond to iodine supplementation or treatment of the underlying hypothyroidism. In cases that do not respond, surgical removal of the gland may be required although this invariably necessitates life-long hormone replacement therapy.

## Summary

- The thyroid gland, and its secretions, are not essential for life, but they are essential for normal development and physical and mental well-being. The hormones secreted by the thyroid gland are tetraiodothyronine, triiodothyronine and calcitonin. The actions of calcitonin are described in Chapter 5.

- The synthesis of the thyroid hormones from tyrosine and iodine begins in the foetus at about the third month of gestation, and continues throughout life. The activity of the thyroid gland is controlled by the hypothalamic–pituitary axis by means of thyrotrophin releasing hormone and thyroid stimulating hormone. The secretion of these hormones is, in turn, controlled by negative feedback. Once secreted, the thyroid hormones are carried within the blood, bound to plasma proteins. It is only the unbound hormone which is biologically active, thus any factor which alters plasma protein binding may influence thyroid hormone activity. In the target tissue most of the tetraiodothyronine is converted to triiodothyronine; the triiodothyronine then produces its effects by interaction with nuclear receptors.

- The main actions of the thyroid hormones are to increase basal metabolic rate and to control growth and development. Thus deficiency of thyroid hormones results in retarded growth, dwarfism, and delayed puberty.

In adults, thyroid hormone deficiency causes decreased basal metabolic rate which is manifested as lethargy and weight gain. Excess thyroid hormones can cause weight loss despite increased appetite, increased heart and respiration rate and insomnia. Both excess and deficient secretion of the thyroid hormones may be accompanied by an enlargement of the thyroid gland: goitre.

## Selected reading

Binkley, S.A. 1995, T$_3$ from the thyroid, immunology, and the thymus. In Binkley, S.A., *Endocrinology*, New York: HarperCollins Publishers, 170–193

Griffin, J.E., 1992), The thyroid. In Griffin, J.E. and Ojede, S.R. (eds), *Textbook of Endocrine Physiology*, 2nd edn, Oxford: Oxford University Press, 224–246

Laycock, J. and Wise, P., 1996, The thyroid. In Laycock, J. and Wise, P., *Essential Endocrinology*, 3rd edn, Oxford: Oxford University Press, 203–240

# 5 Hormonal Control of Calcium Homoeostasis

## 5.1 Introduction

Calcium ions are involved in many vital physiological processes; for example the release of several neurotransmitters and hormones is calcium dependent, the activity of various blood clotting factors requires calcium, and muscle contraction involves calcium. The intracellular and extracellular concentrations of calcium are also integrally involved in determining the transmembrane potential, and therefore in influencing transmembrane sodium and potassium flux and thus the activity of excitable tissues such as nerves. Calcium is also of major skeletal importance as a result of its role in bone. Because of this vital importance, it is necessary to maintain the plasma concentrations of calcium within strict limits, the normal concentration being within the range 2.3–2.6 mmol/l. Of this plasma calcium, approximately 50 per cent is bound to plasma proteins such as albumin or in the form of salts such as citrate or lactate, and approximately 50 per cent is free (ionized). It is only the free calcium which is biologically active, although there is a dynamic equilibrium between the calcium that is bound and that which is free. Plasma calcium, however, only represents a minute portion (less than 1 per cent) of the total body calcium. Approximately 99 per cent (1 kilogram) of the total calcium of the body is in the form of bone, although 1 per cent of that is freely exchangeable with the calcium in other physiological compartments. It is by use of this exchangeable portion of bone calcium that the plasma concentrations are maintained within the strict levels, thus to some extent bone is used as a storage form of calcium.

In children the annual turnover of calcium is greater than the total bone content; in adults it is only 18 per cent of total bone content.

Under normal circumstances there is a constant breakdown, remodelling and reformation of bone. In children there is turnover of more than the total bone content of calcium each year, but in adults there is turnover of only 18 per cent of the bone calcium per year. Bone is covered by a tough periosteum which consists of connective tissue fibres, among which are osteoblasts which synthesize and release molecules of collagen. Calcium and phosphate ions bind specifically to these collagen fibres and form an extracellular matrix around the osteoblast. Dependent upon the relative concentrations of calcium and phosphate ions, calcium hydroxyapatite $(Ca_{10}(PO_4)_6(OH)_2)$, precipitates within the collagen matrix. After calcification of the collagen matrix the cells are referred to as osteocytes. Osteoblasts enhance the precipitation of calcium apatite by causing a local alkalinization resulting in decreased calcium and phosphate ionization. Demineralization of bone is achieved by osteoclasts which release acid phosphatase and hyaluronic acid. The resultant decrease in the local pH causes ionization of the calcium hydroxyapatite and a liberation of calcium. The only other mechanisms by which calcium concentrations can be altered are by variation of intake of calcium through dietary manipulation or an adjustment in the absorption of calcium from the gut or by variation of calcium excretion. All of these factors are controlled by the interplay of several components of the endocrine system, the most important hormones being calcitonin, 1,25-dihydroxychole-calciferol and parathyroid hormone.

## 5.2 Calcitonin

Calcitonin is a peptide hormone, comprising 32 amino acids, which is synthesized in the parafollicular or C cells of the thyroid gland (see Chapter 4). Like other peptide hormones, calcitonin is initially synthesized and stored as a large precursor molecule. The secretion of calcitonin is influenced directly by plasma calcium concentrations, thus increased plasma calcium causes a parallel increase in calcitonin secretion. The gastrointestinal hormone gastrin has also been shown to stimulate calcitonin secretion, but this process is probably insignificant under normal circumstances. In some cases of Zollinger–Ellison syndrome, in which there is abnormally high secretion of gastrin, there is an elevation of circulating calcitonin. The calcitonin secretion in response to raised plasma calcium is greater in males than in females, although in both sexes the response decreases with age. These sex and age differences may be related to changes in the secretion of the oestrogenic and andro-genic sex hormones (see Chapters 10 and 11), which have been shown to enhance calcitonin secretion.

Calcitonin acts on membrane bound $G_2$-protein linked receptors on the osteoclasts and in the kidney. It is believed that cAMP is the second messenger. Calcitonin inhibits the actions of osteoclasts and therefore prevents the mobilization of calcium and phosphate from bone. It also acts on receptors on the ascending limb of the loop of Henlé and the distal tubule of the nephron to decrease reabsorption of calcium and phosphate; there is also an accompanying decrease in sodium and potassium reabsorption. These actions therefore have the effect of lowering plasma calcium concentrations (see Figure 5.1). One interesting feature of the effect of calcitonin is the 'escape phenomenon' in which the effect of the calcitonin appears to be self-limiting. It has been suggested that this effect may be related to receptor down regulation (see Chapter 1).

The gene responsible for calcitonin secretion is also expressed in nervous tissue, particularly the hypothalamus and pituitary, where it induces the synthesis of a 37 amino acid peptide called calcitonin gene related peptide (CGRP) which may have actions as a neuro-modulator. CGRP is also found in the thyroid gland.

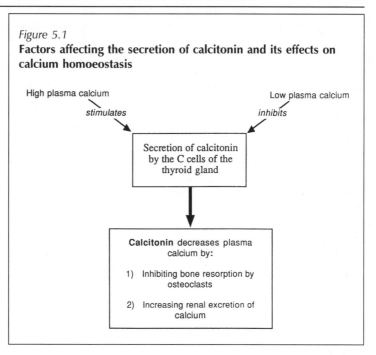

Figure 5.1

**Factors affecting the secretion of calcitonin and its effects on calcium homoeostasis**

High plasma calcium

*stimulates*

Low plasma calcium

*inhibits*

Secretion of calcitonin by the C cells of the thyroid gland

**Calcitonin** decreases plasma calcium by:

1) Inhibiting bone resorption by osteoclasts

2) Increasing renal excretion of calcium

## 5.3 1,25-Dihydroxycholecalciferol

1,25-Dihydroxychole-
calciferol is related to
vitamin D.

The steroid derivative 1,25-dihydroxycholecalciferol is related to vitamin D₃, cholecalciferol. Cholecalciferol itself is inactive, and therefore it has been suggested that its labelling as a vitamin is incorrect, the definition of a vitamin being 'an organic dietary constituent necessary for life and development but which does not act as a dietary energy source'. Cholecalciferol is obtained from the diet and is converted to 25-hydroxycholecalciferol in the liver. This metabolite must then be further converted to 1,25-dihydroxychole-calciferol (1,25-DHCC) in the kidney. It is this latter conversion, which occurs exclusively in the kidney, that renders 1,25-DHCC suitable for relabelling as a hormone: a substance secreted by a gland (in this case the kidney) and transported by the blood to the target tissues. The kidney is also able to convert the 25-hydroxycholecalciferol to 24,25 dihydroxycholecalciferol (24,25-DHCC), which is biologically inactive (see later) Within the blood 1,25-DHCC and 24,25-DHCC are transported either within chylomicrons or bound to a plasma protein, calciferol binding globulin.

The release of 1,25-DHCC varies inversely with plasma calcium concentrations, but there is no evidence that calcium itself can influence the synthesis of the hormone. The regulation appears to be via effects of parathyroid hormone (see later), 1,25-DHCC itself and the influence of phosphate ions on the activity of the enzyme required for 1,25-DHCC synthesis. Thus in the presence of low plasma calcium concentrations there is an increase in the secretion of parathyroid hormone. One effect of parathyroid hormone is to increase the synthesis of the biologically active 1,25-DHCC and to decrease the synthesis of the inactive 24,25-DHCC. At times of increased calcium mobilization from bone there is an accompanying increase in plasma phosphate concentration; it is the raised phosphate ion concentration which increases the synthesis of the inactive 24,25-DHCC and decreases the release of 1,25-DHCC. Calcitonin will therefore influence 1,25-DHCC release indirectly by influencing plasma calcium and phosphate concentrations; 1,25-DHCC directly inhibits the enzyme responsible for its production and thus enhances the release of 24,25-DHCC (see Figure 5.2).

Like steroid hormones, 1,25-dihydroxychole-calciferol crosses the cell membrane and acts on intracellular receptors.

The structure of 1,25-DHCC resembles that of a steroid, as does its mechanism of action (see Chapter 1). Like the steroid hormones, 1,25-DHCC crosses the cell membrane and acts on specific receptors within the cytoplasm and nucleus to influence protein synthesis. The effects of 1,25-DHCC are not seen until two hours after its interaction with its receptors because of this requirement for protein synthesis. One of the proteins synthesized is calcium binding protein (CaBP) which enhances the movement of calcium ions from the brush border of intestinal cells into the cytoplasm. Receptors for 1,25-DHCC have been identified in a variety of cells including intestinal cells, osteoblasts and the distal tubule of the kidney. The major effect of 1,25-DHCC is to raise plasma calcium concentrations. In the kidney reabsorption of calcium by the distal tubule and reabsorption of both calcium and phosphate in the proximal

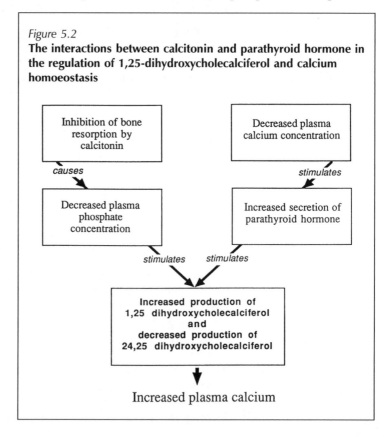

Figure 5.2

**The interactions between calcitonin and parathyroid hormone in the regulation of 1,25-dihydroxycholecalciferol and calcium homoeostasis**

tubule is enhanced and in the gut there is increased uptake of both calcium and phosphate, by different processes (see Figure 5.3). This latter action mediates the major effect of 1,25-DHCC on calcium homoeostasis. Paradoxically, 1,25-DHCC increases the activity of osteoblasts and therefore promotes the synthesis of collagen and the incorporation of calcium.

## 5.4 Parathyroid hormone

The other hormone which plays a major role in calcium homoeostasis is parathyroid hormone (parathormone, PTH) which is secreted by the parathyroid glands. There are commonly four parathyroid glands, all of which are located close to the posterior surface of the thyroid gland, embedded at the upper and lower margins of each thyroid lobe. Each parathyroid gland is relatively small and is surrounded by a fibrous capsule. The glands contain two principal types of epithelial cell: the chief cells which are responsible for the secretion of PTH, and the oxyphil cells which appear at puberty, but for which there is no known function. At any one time only about 25 per cent of the chief cells are actively synthesizing and storing PTH; the remainder are inactive.

PTH is a large, single chain peptide hormone with 84 amino acids. Like other peptide hormones, PTH is initially synthesized as a large precursor molecule, preproparathyroid hormone (prepro-PTH), which contains 115 amino acids. This is converted to

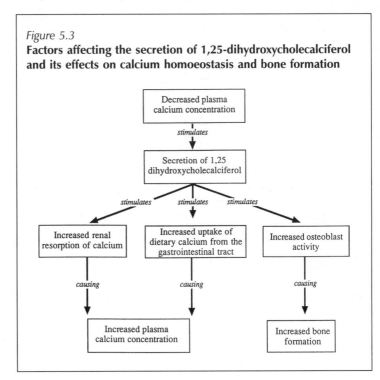

*Figure 5.3*

**Factors affecting the secretion of 1,25-dihydroxycholecalciferol and its effects on calcium homoeostasis and bone formation**

proparathyroid hormone (pro-PTH), which contains 90 amino acids, and thence to PTH which is secreted. Pro-PTH is normally converted to PTH within 20 minutes of its synthesis, thus the parathyroid gland normally contains very small amounts of pro-PTH. Following stimulation of PTH secretion, the hormonal release occurs in two phases: an initial, rapid release of stored PTH followed by a delayed release of newly synthesized hormone. Within the circulation PTH is cleaved to form two major peptide fragments, one of which retains the biological activity of the hormone, the 34 amino acids of the amino terminal. This cleavage takes place within the liver, the kidneys and within the chief cells of the parathyroid gland. PTH has a plasma half-life of three to four minutes.

Secretion of PTH normally involves the production of cAMP, and is thus enhanced by stimulation of $\beta$-adrenoceptors on the surface of the chief cells. It also appears to be a magnesium dependent process as hypomagnesaemia is accompanied by decreased PTH secretion while abnormally high plasma magnesium enhances secretion. The most important factor in the control of PTH secretion, however, is the direct effect of plasma calcium: raised plasma calcium decreases PTH secretion. This effect is probably mediated by an inhibition of cAMP production by calcium although it is possible that calcium may influence PTH secretion by an effect on ion channels within the membrane of the chief cell. For example, activation of a calcium dependent potassium channel may lead to hyperpolarization of the cell membrane, and thus decreased secretion of PTH. Both phases of PTH secretion are inhibited by raised calcium. 1,25-DHCC also decreases PTH secretion possibly by direct inhibition of hormone synthesis or secondarily to its actions on plasma calcium.

PTH acts to increase plasma calcium and decrease plasma phosphate concentrations. PTH acts on $G_s$-protein linked, membrane bound receptors in bone and in the kidney to cause production of the second messenger cAMP. In some cells, however, the effect of the PTH is to stimulate production of the second messengers $IP_2$ and DAG (see Chapter 1). Two models have been proposed for this diversity in the mechanism of action: one model suggests that there is a single PTH receptor which possess two transduction mechanisms, the other model suggests that there are two separate types of PTH receptor, one linked to adenyl cyclase and the other linked to phospholipase C. This matter remains unresolved.

Two models have been proposed for the actions of parathyroid hormone (PTH). One proposes a single PTH with two separate transduction pathways, the other proposes two separate PTH receptors.

In bone PTH acts on receptors on osteoblasts to inhibit bone formation. PTH also stimulates osteoclast activity, stimulating those enzymes required for bone resorption, although there is no evidence of osteoclasts expressing PTH receptors. This anomaly has been overcome by the suggestion that PTH causes the production of an osteoclast stimulating factor. It is believed that 1,25-DHCC has a permissive effect on the actions of PTH in bone, possibly as a result of its effects on protein synthesis (see Figure 5.4).

In the kidney PTH decreases reabsorption of phosphate ions and increases reabsorption of calcium ions in the proximal tubule and, to a lesser extent, in the distal tubule. It also increases the excretion

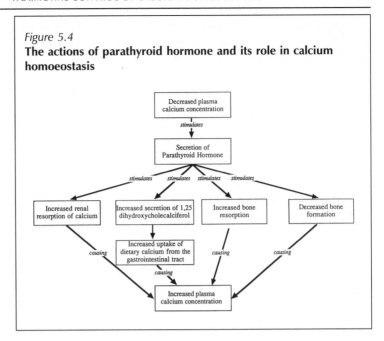

*Figure 5.4*
**The actions of parathyroid hormone and its role in calcium homoeostasis**

of sodium, potassium and hydrogen carbonate ions and decreases the excretion of magnesium, ammonium and hydrogen ions. The resultant acidosis favours the dissociation of calcium from bone and from plasma proteins. These effects on the kidney therefore lead to an increase in the concentration of free calcium in the plasma. The other important action of PTH in the kidney is the stimulation of 1,25-DHCC synthesis (see above).

In the gastrointestinal tract PTH enhances uptake of both calcium and phosphate, however this effect is mediated by 1,25-DHCC. The net effects of PTH are therefore: to liberate calcium and phosphate from bone by inhibiting the formation of new bone and by stimulating bone resorption; to enhance uptake of calcium and phosphate in the gastrointestinal tract; to act on the kidney to increase phosphate excretion and calcium reabsorption; and to promote an acid environment which increases the proportion of free calcium.

## 5.5 Integration of the hormonal control of calcium metabolism

Blood concentrations of total calcium are maintained at a fairly constant value of approximately 2.5 mmol/l, although there are normal diurnal fluctuations of approximately 5 per cent, with a peak at about midday and a nadir at about 0200. The pattern for the variation of free calcium, however, is very different with a peak at about 0400 and a nadir at about 1700. These variations probably mirror the pattern of dietary intake of calcium and proteins and the subsequent variation of plasma protein concentration. As stated earlier, because of its role in a wide range of physiological processes, it is necessary to maintain calcium levels within narrow limits, but it is also important to maintain the integrity of the bone structure. It is this requirement to manage calcium homoeostasis within narrow limits without compromising skeletal integrity that has led to the development a complex three hormone control system.

Parathyroid hormone increases plasma calcium concentrations by a variety of mechanisms, but in doing so it decreases the formation of new bone and increases bone resorption, thereby compromising sketetal integrity. On the other hand, 1,25-DHCC increases plasma calcium without having the deleterious effects on bone, it actually promotes bone formation by osteoblasts and does not increase the activity of osteoclasts. Thus by varying the relative activity of these hormones, plasma calcium concentration and bone turnover can be controlled independently. The role of calcitonin is to assist with the homoeostasis of plasma calcium while not having a profound effect on bone formation and integrity. In the light of the interrelationship between these hormones, and the fact that the primary symptoms of disorders of PTH secretion are alterations in plasma calcium concentration, while deficiency of 1,25-DHCC causes bone deformities (see later) it is tempting to suggest that the primary role of 1,25-DHCC is to control bone turnover, while the actions of PTH and calcitonin regulate plasma calcium, taking into account the actions of 1,25-DHCC.

> The primary role of 1,25-DHCC is the control of bone turnover. The role of calcitonin and PTH is the control of plasma calcium concentrations, taking into account the actions of 1,25-DHCC.

## 5.6 Disorders of endocrine control of calcium metabolism and their treatment

### 5.6.1 *Hyperparathyroidism*

Excess secretion of PTH occurs in approximately 0.02–0.1 per cent of the population, being twice as common in females as in males. The condition usually presents between the ages of 30 and 60 years. In 85 per cent of cases it is caused by a tumour of a single parathyroid gland, but in a minority of cases there may be involvement of two or more of the glands. This condition is referred to as primary hyperparathyroidism. Secondary hyperparathyroidism is the condition in which the elevated secretion of PTH is a consequence of a decrease in plasma calcium brought about by another disorder, for example kidney failure. In some cases a deficiency in vitamin D (or more correctly 1,25-DHCC) results in lowering of plasma calcium and thus an increased secretion of PTH; initially this compensates for the lack of 1,25-DHCC, but eventually the parathyroid glands become enlarged and the secretion of PTH becomes excessive. This condition is referred to as tertiary hyperparathyroidism.

Hyperparathyroidism may also occur as part of a group of disorders called multiple endocrine neoplasias (MEN). In these conditions, tumours arise within several endocrine glands simultaneously, although the presenting symptoms may initially only indicate the involvement of one gland. The various forms of MEN are

> Hyperparathyroidism may occur as a feature of one of the multiple endocrine neoplasias (MEN).

known to be inherited disorders and the involvement of several glands probably arises because of their common embryological origin (e.g. APUD cells, see Chapter 6). In MEN type I, which affects 1 in 5000 of the population, the most common feature is hyperparathyroidism, although the patients also develop tumours of the pancreas and the anterior pituitary gland. MEN IIa is also characterized by hyperparathyroidism in 20 per cent of cases, although the presentation with tumours of the adrenal medulla or the thyroid gland is more common. Several tumours, for example carcinoma of the prostate gland, cause hypercalcaemia due to the production of PTH related peptide. This is an 139 amino acid peptide with actions similar to PTH. Its presence therefore gives rise to symptoms which may be misdiagnosed as hyperparathyroidism.

Hyperparathyroidism is usually diagnosed after a long period of illness, during investigations of other conditions. Its predominant symptoms are those caused by the hypercalcaemia, thus there is tiredness, lethargy and general feelings of being unwell; there may also be polyuria with dehydration. If untreated the condition may progress to an increase in muscle weakness and cardiac arrhythmias. There may also be mental confusion which may culminate in coma or death. In addition to these features there may be renal calculi and occasionally evidence of bone rarefraction. This latter condition often results in bone pain, especially in secondary and tertiary hyperparathyroidism where there may be a concomitant deficiency of 1,25-DHCC.

> The predominant symptoms of hyperparathyroidism are tiredness, lethargy and general feeling of being unwell.

The treatment of primary and tertiary hyperparathyroidism is usually by surgical removal of the parathyroid glands followed by treatment with calcium and 1,25-DHCC. Alternative approaches are the use of drugs such as the diuretic frusemide to potentiate excretion of calcium, the administration of one of a group of drugs called biphosphonates which directly inhibit bone resorption, or administration of calcitonin to counteract the effects of PTH. Until recently the only available source of calcitonin was a chemically synthesized version of salmon calcitonin (see Box 5.1), but in early 1997 salmon calcitonin produced by recombinant DNA technology was developed. This can be produced in much greater quantities and at much lower cost than its synthetic equivalent. In secondary hyperparathyroidism it is the underlying disorder that is treated, for example in the case of hyperthyroidism associated with end stage renal disease a recently developed product, one-alpha D2, has been shown to reduce plasma parathyroid hormone concentrations by up to 90 per cent within 12 weeks of starting treatment.

### 5.6.2  *Hypoparathyroidism*

Removal of the parathyroid glands results in death within hours, however a relative deficiency of PTH is often symptomless, or may be limited to paraesthesia of the fingers and toes. With time, however the condition may lead to the deposition of calcium in nervous tissue or the lens of the eye (cataracts) and the increased excitability of nerves and muscle which may lead to tetany or convulsions. This condition may be caused by an autoimmune destruction of the parathyroid glands or the removal of or damage to the glands during thyroidectomy. In pseudohypoparathyroidism the secretion of PTH by the gland is normal, however a genetic defect of the PTH receptor results in a failure to generate the cAMP second messenger. The symptoms of this condition normally become apparent during early life and include mental deficiency, epilepsy, tetany, bone deformation and shortness of stature. Treatment of this hypoparathyroidism is by calcium supplementation together with replacement therapy with either PTH, which is expensive, or a 1,25–DHCC analogue; obviously the former is inappropriate for the treatment of pseudohypoparathyroidism (see Box 5.1).

### 5.6.3.  *Osteomalacia*

In children rickets can be due to inadequate intake of vitamin D, or inability of the skin to synthesize vitamin D due to lack of exposure to sunlight.

Osteomalacia (which in children is referred to as rickets) is caused by a deficiency of 1,25-DHCC. The condition may be caused by an inadequate dietary intake of vitamin D, the inability of the skin to synthesize vitamin D, for example due to the lack of exposure to ultraviolet radiation, or an inability to convert the vitamin to the active metabolite, due to liver disease for example. The most important feature of this condition is the failure of bone development in children, leading to the bowing of the long, load bearing bones of the legs. In adults the strength of the bones becomes compromised because of abnormal calcification and there is bone pain. The features of osteomalacia differ from those of hyperparathyroidism because the 1,25-DHCC deficiency results in a decrease in osteoblast function, while in hyperparathyroidism the bone rarefraction is a result of increased osteoclast function. The treatment of osteomalacia is by supplementation with 1,25-DHCC or a related compound.

### 5.6.4.  *Osteoporosis*

A gradual loss of bone mass after the age of 30 is an inevitable consequence of aging.

Osteoporosis is a condition in which there is a loss, not only of the calcium moiety of bone but also the protein moiety. Because of this there is considerable weakening of the bone and an increased incidence of fractures. A gradual loss of bone mass after the age of 30 is an inevitable effect of aging, however in most males the condition does not become manifest until very late in life, if at all. In females, however, the cessation of ovarian function at the menopause, and the subsequent decrease in circulating oestrogens results in an acceleration of the osteoporotic process. The progression of osteoporosis can be limited by oestrogen replacement therapy postmenopausally, but restoration of existing bone loss is very rarely achieved (see Chapter 11).

Unlike the osteoporosis associated with the menopause where the condition is a result of sex hormone deficiency, prolonged (6–12 month) treatment with corticosteroids may result in osteoporosis due to the decreased protein synthesis and increased mobilization of the protein matrix for use in gluconeogenesis (see Chapter 9).

Box 5.1  **Drugs affecting bone and calcium metabolism**

Calcitonin or one of its analogues may be used to lower plasma calcium concentrations, for example in patients with hyperparathyroidism or pseudohyperparathyroidism associated with malignant disease:

| Drug | Proprietary name |
|------|------------------|
| Calcitonin (porcine) | Calcitare (injection) |
| Calcitonin (salmon) | Calsynar (injection) |
| Calcitonin (salmon) | Miacalcic (injection) |

Biphosphonate drugs combine with the hydroxyapatite component of bone and slows its turnover. Such drugs are therefore used in the treatment of disorders involving bone resorption, for example Paget's disease and post-menopausal osteoporosis:

| Drug | Proprietary name |
|------|------------------|
| Alendronic acid | Fosamax (tablets) |
| Disodium etidronate | Didronel (tablets) |
| Disodium pamidronate | Aredia dry powder (injection) |
| Sodium clodronate | Bonefos (tablets, capsules, injection) |
| Sodium clodronate | Loron (injection) |
| Tiludronic acid | Skelid (tablets) |

Osteoporosis may also develop, particularly in young people, following extended periods of immobilization, or weightlessness.

Attempts have been made to reverse the breakdown of the bone matrix by administration of calcitonin and anabolic steroids, but, as stated above, osteoporosis does not respond well to treatment. Therapies aimed at its prevention, however, have been very successful. Use of post-menopausal oestrogen replacement can delay the onset of the condition in females almost indefinitely, and there is evidence that regular physical exercise coupled with calcium supplements may decrease the risk of developing the condition. Co-administration of biphosphonates, which inhibit bone resorption, has also been shown to reduce the risk of bone loss during long-term corticosteroid therapy.

### 5.6.5 *Paget's disease*

This is a condition in which there is an increase in the activity of both osteoblasts and osteoclasts. It occurs in up to 10 per cent of the population over the age of 60 but the cause of the condition is unknown, although it is not believed to be endocrine in origin. The activity of the osteoclasts predominates in the early stages of the disease which is characterized by weakening and deformity of the bones. In the later stages the activity of the osteoblasts may predominate, leading to bone thickening and bone pain. To date the treatments have been directed only towards the early phase with the use of calcitonin and the biphosphonates to decrease osteoclast activity. In general the treatments do not bring full relief from the condition.

# Summary

- Calcium ions are vital for physiological processes such as release of neurotransmitters and blood clotting; they are also of major importance as part of the structure of bone. Calcium homoeostasis is achieved by interaction of three hormones: calcitonin, 1,25-dihydroxycholecalciferol (1,25 DHCC) and parathyroid hormone.

- Calcitonin, a peptide hormone secreted by the C cells of the thyroid gland, is released at times of raised plasma calcium. It acts to decrease calcium concentration by inhibiting bone reaborption and by increasing calcium excretion by the kidney.

- The steroid-like 1,25-DHCC is synthesized in the kidney by the conversion of vitamin D. The synthesis of 1,25-DHCC is inversely related to plasma calcium concentration, thus it is synthesized primarily at times of low plasma calcium. It acts to raise plasma calcium concentrations by enhancing calcium reabsorption in the kidney and increasing gut uptake of dietary calcium. Paradoxically, 1,25-DHCC increases the rate of bone formation.

- Parathyroid hormone is a peptide hormone secreted by the parathyroid glands in response to decreased plasma calcium. Parathyroid hormone raises plasma calcium concentrations by stimulating 1,25-DHCC synthesis, inhibiting formation of new bone and enhancing bone resorption.

- Hyperparathyroidism results in hypercalcaemia, the symptoms of which are tiredness, lethargy and muscle weakness. The condition can be treated by administration of calcitonin to lower the calcium concentrations. Hypoparathyroidism results in hypocalcaemia which can cause tetany or convulsions. This condition can be treated by administration of 1,25 DHCC. A deficiency of 1,25 DHCC results in rickets or osteomalacia.

## Selected reading

Breslau N.A., 1992, Calcium homeostasis. In Griffin, J.E. and Ojede, S.R. (eds), *Textbook of Endocrine Physiology*, 2nd edn, Oxford: Oxford University Press, 276–308

Hadley, M.E, 1992, Hormonal control of calcium homeostasis. In Hadley, M.E., *Endocrinology*, 3rd edn, Englewood Cliffs: Prentice-Hall, 208–238

Laycock, J. and Wise, P., 1996, Calcium regulation, bone, and its metabolic disorders. In Laycock, J. and Wise, P., *Essential Endocrinology*, 3rd edn, Oxford: Oxford University Press, 241–273

# 6 Hormonal Control of the Gastrointestinal Tract

## 6.1 Introduction

As stated in Chapter 1, a definition of the endocrine system would be a system in which a group of cells (a gland) secretes a potent transmitter substance (a hormone) into the blood. The transmitter is then carried by the blood to the target cells where a response is elicited. In this aspect the endocrine system of the gastrointestinal tract differs from that of other areas of the body, with the possible exception of the brain. Whereas normally, hormones are secreted by a discrete group of cells (a gland), in the gut the hormones are produced by cells that are scattered diffusely within the gastric and intestinal mucosa. Secondly, the gut hormones may not enter the general circulation. Following their secretion they may be carried in the blood for only a short distance before they reach their target cells where they are degraded after they have elicited their response, although many gut hormones can be detected in the blood. A similar situation is encountered in the brain where neurohormones are secreted by diffusely distributed neurones and, in the case of the hypothalamic–pituitary axis, the hormones are transported directly to their target by the hypothalamicoadeno-hypophyseal portal system. Most neurohormones cannot be detected in the systemic circulation.

The endocrine system of the gastrointestinal tract differs from that of any other system of the body, with the exception of the brain.

This similarity between the brain and the gut is not coincidental as it is suggested that some of the cells of the gut are derived from the same embryological origins as the brain. The cells in question have been labelled the APUD cells, the acronym being derived from their abilities to form amines (amine precursor uptake and decarboxylation). APUD cells of the gut are believed to derive from the ectodermal neural crest of the embryo, which eventually forms the brain and spinal cord. It is because of this that there is a great deal of commonality in the physiology and biochemistry of the gastrointestinal tract and the brain. A prime example of this similarity is the fact that both the brain and the gut possess receptors for 5-hydroxytryptamine (5-HT, serotonin), acetylcholine and noradrenaline. In the brain these substances act as neurotransmitters. In the gut, 5-HT and acetylcholine induce smooth muscle contraction, and in noradrenaline causes smooth muscle relaxation. This parallelism persists into the arena of endocrinology where several hormones have been identified in both the gut and the brain. In the gut they are involved in the regulation of peristalsis and the secretion of gastrointestinal juices while in the brain the same chemicals play

The similarity between the transmitters of the brain and the gut may reflect their common embryological origin.

a role in the modulation of neuronal activity. In most cases the actions of the hormones are highly localized thus their functions at the two sites are independent. At times of aberrant secretion however, actions at both sites may become apparent; the same delocalization of effects may occur following attempts to influence the hormonal action at either of the sites by the administration of drugs. It must also be remembered that within the embryo, several other endocrine glands, for example the thyroid gland and the anterior pituitary gland, are derived from the gut tissue. Thus in some cases, disorders which affect the APUD cells of the gastrointestinal tract and the brain are also manifest as disorders of these other glands.

## 6.2 Structure and function of the gastrointestinal tract

Before considering the actions and functions of gut hormones it is necessary to review briefly the processes that occur as food passes along the gastrointestinal tract. Within the mouth the food is masticated and mixed with saliva. The saliva serves partly as a lubricant but because it contains enzymes such as $\alpha$-amylase it also begins the process of carbohydrate digestion. From the mouth the food passes down the oesophagus into the stomach which contains gastric secretions comprised of not only hydrochloric acid but also pepsins. In the presence of hydrochloric acid the activity of the salivary amylase is inhibited, but, conversely, the actions of the pepsins are activated; this leads to the hydrolysis of peptide bonds. The gastric contents, chyme, next enter the duodenum. Within the duodenum the activity of the gastric pepsins is terminated by the increased alkalinity, while the enzymes secreted by the pancreas further digest the polysaccharides and proteins; the addition of bile salts in the duodenum facilitates the emulsification of fats. The food is then propelled along the gastrointestinal tract by rhythmic contractions of the longitudinal and circular smooth muscles which make up the gut wall. This peristaltic activity is controlled predominantly by the activity of the sympathetic and parasympathetic branches of the autonomic nervous system. As a result of the peristaltic activity the food leaves the duodenum and enters the jejunum and the ileum where further digestion of foodstuffs occurs followed by absorption of sugars, amino acids and fatty acids into the bloodstream. The final stages of absorption occur in the colon where the remaining gut contents are dehydrated prior to evacuation via the rectum.

> The activity of the gut must be highly co-ordinated in order to prevent erosion of the intestinal mucosa by the digestive enzymes.

An important feature of digestion and the activity of the gut is that it must be highly co-ordinated. If, in the absence of sufficient foodstuffs, there is secretion of gastric acid and pepsin in the stomach, the enzymes and acid will begin to attack the gastric mucosa resulting in gastric ulceration. Similarly, if the contents of the stomach enter the duodenum before the pancreatic secretions are released, the acidity of the chyme results in duodenal damage. In the colon, if there is an increased rate of peristalsis it results in insufficient removal of water from the gut contents, which produces diarrhoea; decreased peristalsis culminates in constipation. It is because of this need for co-ordination that the integrated gastrointestinal endocrine system is so important.

## 6.3 Gastrointestinal hormones

The gastrointestinal hormones are synthesized by a group of cells known variously as the clear cells, enterochromaffin cells, argentaffin cells or agyrophil cells. These cells are distributed along the whole length of the gut, although their density varies between regions; it is because of the diffuse distribution of these hormone secreting cells that the system is sometimes referred to as the dispersed endocrine system (DES). It has been proposed that the benefit of such a diffuse system is that the cells are able to repond to stimuli (i.e. the presence of metabolic products) that may arise at many points along the gut, rather than having just one locus of stimulus receptivity and many of the hormone secreting cells have areas of their membranes directly adjacent to the gut lumen.

Based upon their chemical structures the gastrointestinal hormones can be divided into three groups. The first group is the gastrin–cholecystokinin family which contains gastrin and the cholecystokinins; the second group is the secretin family which contains secretin, glucagon, vasoactive intestinal peptide and gastric inhibitory peptide; the final group consists of peptides such as somatostatin, motilin and substance P. The synthesis, structure actions of each of these hormones, will be covered individually prior to a description of how they function within the gut in a co-ordinated fashion.

### 6.3.1  *The gastrin–cholecystokinin family of gastrointestinal hormones*

The physiologically important form of gastrin contains 17 amino acids while cholecystokinin contains 33 amino acids; the two hormones are grouped together however because they possess an identical sequence of the terminal five amino acids at the C-terminal (see Table 6.1).

### Gastrin

Several forms of gastrin can be found although the predominant form in the circulation is comprised of 34 amino acids (G34). A larger form, G45, is believed to be a precursor of the other gastrins. The form with greatest physiological significance is probably G17; it is possible that G14 is a breakdown product. Gastrin (G17) is synthesized in the antral portion of the stomach by the G cells; in the duodenum gastrin is synthesized as G34. The secretion of gastrin is stimulated by the presence of food within the stomach, most notably peptides or amino acids and to a lesser extent fatty acids; stomach distension also results in gastrin secretion. In the absence

**Table 6.1** A comparison of the chemical structures of gastrin and cholecystokinin (note commonality of the terminal amino acid residues)

| Gastrin (34 amino acids) | Cholecystokinin (39 amino acids) |
|---|---|
| | tyr- |
| | ile- |
| | gln- |
| | gln- |
| pyro- | ala- |
| glu- | arg- |
| leu- | lys- |
| gly- | ala- |
| **pro-** | **pro-** |
| gln- | ser- |
| **gly-** | **gly-** |
| pro- | arg- |
| pro- | met- |
| his- | ser- |
| leu- | ile- |
| **val-** | **val-** |
| ala- | lys- |
| asp- | asn- |
| pro- | leu- |
| ser- | gln- |
| lys- | asn- |
| lsy- | leu- |
| gln- | asp- |
| gly- | pro- |
| pro- | ser- |
| trp- | his- |
| leu- | arg- |
| glu- | ile- |
| glu- | ser- |
| glu- | asp- |
| glu- | arg- |
| glu- | asp- |
| ala- | tys- |
| tys- | met- |
| **gly-** | **gly-** |
| **trp-** | **trp-** |
| **met-** | **met-** |
| **asp-** | **asp-** |
| **phe-** | **phe-** |
| **NH$_2$** | **NH$_2$** |

of food, secretion can also be stimulated by increased activity of the autonomic nervous system: both adrenaline and stimulation of the vagus nerve result in gastrin secretion. The vagal neurotransmitter reponsible for this effect is probably gastrin releasing peptide (GRP).

The main action of gastrin is to increase the secretion of gastric acid. This is via a direct effect on the parietal cells or it can involve the potentiation of histamine induced gastric acid secretion. As a result of the increased acid secretion, gastrin increases pepsin activity. Gastrin also increases the secretion of pancreatic enzymes in readiness for the movement of the stomach contents into the duodenum, and the secretion of the pancreatic hormones insulin and glucagon (Chapter 7). It is gastrin which facilitates the movement of the stomach contents into the duodenum by increasing both gastric and intestinal motility and by relaxing the pyloric sphincter; it also stimulates contraction of the gallbladder which results in the addition of bile salts to the gut contents. Another important function of gastrin is its role as a growth factor. Gastrin stimulates the growth of the gastric mucosa and in animals in which the gastric antrum has been removed there is atrophy of the gastric and intestinal mucosa and the pancreas due to removal of the source of the hormone. The secretion of gastrin is under feedback control and is inhibited by an increased acid concentration within the stomach and by other gastrointestinal hormones such as vasoactive intestinal peptide, somatostatin and glucagon.

Gastrin has also been identified in the brain, in the G17 and G34 forms, and administration of the terminal four amino acids has been shown to stimulate secretion of growth hormone by the anterior pituitary gland; thus it has been suggested that gastrin may play a role as a neurohormone.

## Cholecystokinin

The most common form of cholecystokinin (CCK) is comprised of 33 amino acids, the terminal five of which are identical to those of gastrin; there are however other forms of CCK, all with the same terminal amino acids. Forms with 8, 12 and 58 amino acids can be found in the I cells of the intestinal tract, while in the blood the predominant forms contain 8, 33 or 58 amino acids. CCK is secreted by the I cells in response to the passage of gastric acid or products of digestion such as amino acids and fatty acids into the duodenum.

First it was pancreozymin, then it was cholecystokinin-pancreozymin, now it is cholecystokinin.

Once secreted, the most notable action of CCK is to cause increased production of hepatic bile and contraction of the gallbladder to cause the addition of bile salts to the duodenal contents. Another important action of CCK is the stimulation of the secretion of pancreatic digestive enzymes, insulin, glucagon and pancreatic polypeptide. Initially the agent responsible for the stimulation of pancreatic secretion was named pancreozymin, but it was later established that pancreozymin and CCK bore the same chemical structure, thus the joint name of cholecystokinin–pancreozymin (CCK–PZ) was adopted. This has now largely been replaced by the single term CCK. Other actions of CCK include inhibition of gastric motility, closure of the pyloric sphincter and stimulation of peristalsis of the duodenum, jejunum, ileum and colon. The closure of the sphincter prevents reflux of duodenal contents into the stomach. Unlike gastrin, CCK has no effect on the growth of the gut mucosa, however it does cause increased growth of the exocrine pancreas. The secretion of CCK ceases when the products of digestion leave the duodenum and enter the jejunum.

CCK is also found in the brain, predominantly in the 8 amino acid form although also in the 58 amino acid form, where it is believed to act as a neurotransmitter or neuromodulator. Several workers have shown that administration of CCK to animals causes a reduction in food intake, thus it has been suggested that CCK is involved in satiety.

### 6.3.2  *The secretin family of gastrointestinal hormones*

Secretin, glucagon, vasoactive intestinal peptide and gastric inhibitory peptide are classed together because of the degree of homology in their amino acid sequences; however, unlike the gastrins and CCK, this homology is not in a distinct portion of the molecules, but is, rather, in diffuse singlets, doublets and triplets of amino acids (see Table 6.2).

### Secretin

Secretin is comprised of 27 amino acids, of which 14 are homologous to the sequence of glucagon. It is secreted by S cells which are found in the duodenum and, to a lesser extent, in the jejunum and ileum. The stimulus for the secretion of secretin is the presence of gastric acid in the duodenum. Following its secretion, secretin acts via cAMP to stimulate the secretion of a pancreatic juice with a high concentration of bicarbonate ions; an action which possibly involves interaction with CCK. Secretin also increases the secretion of insulin and bile and causes constriction of the pyloric sphincter and a decrease in gastric acid secretion. The secretion of secretin ceases when the pH of the duodenal contents rises above 4.5 as a result of the alkaline pancreatic secretion.

Secretin has also been identified in the brain where its function is unclear.

### Glucagon

Glucagon, which is comprised of 29 amino acids, is secreted by the $\alpha$-cells of the stomach, duodenum and islets of Langerhans. It does not appear to have a direct action on the exocrine secretions of the gastrointestinal tract, nor on gut motility. Its actions are restricted to its effects on carbohydrate metabolism as described in Chapter 7.

### Vasoactive intestinal peptide

Vasoactive intestinal peptide (VIP) contains 28 amino acids, just seven of which are homologous with secretin. It is produced by D cells along the whole length of the gastrointestinal tract, but its secretion is not increased in response to food intake. Within the gut, VIP inhibits gastric acid secretion and stimulates the endocrine and exocrine secretions of the pancreas; however it also has other

**Table 6.2**  A comparison of the chemical structures of the secretin family of gastrointestinal hormones (note the commonality of many of the amino acid residues)

| Glucagon (29 amino acids) | Secretin (27 amino acids) | Vasoactive intestinal peptide (28 amino acids) | Gastric inhibitory peptide (42 amino acids) |
|---|---|---|---|
| **his-** | **his-** | **his-** | tyr- |
| **ser-** | **ser-** | **ser-** | ala- |
| gin- | asp- | asp- | glu- |
| **gly-** | **gly-** | ala- | **gly-** |
| **thr-** | **thr-** | val- | **thr-** |
| **phe-** | **phe-** | **phe-** | **phe-** |
| **thr-** | **thr-** | **thr-** | ile- |
| **ser-** | **ser-** | asp- | **ser-** |
| **asp-** | glu- | asn- | **asp-** |
| **tyr-** | leu- | **tyr-** | **tyr-** |
| **ser-** | **ser-** | thy- | **ser-** |
| lys- | **arg-** | **arg-** | ile- |
| tyr- | **leu-** | **leu-** | ala- |
| leu- | **arg-** | **arg-** | met- |
| **asp-** | glu- | lys- | **asp-** |
| ser- | gly- | gln- | lys- |
| arg- | ala- | met- | ile- |
| **arg-** | **arg-** | ala- | his- |
| ala- | leu- | val- | gln- |
| **gln-** | **gln-** | lys- | **gln-** |
| **asp-** | arg- | lys- | **asp-** |
| **phe-** | leu- | tyr- | **phe-** |
| **val-** | **leu-** | **leu-** | **val-** |
| glu- | **gln-** | asn- | asn- |
| **trp-** | gly- | ser- | **trp-** |
| **leu-** | **leu-** | ile- | **leu-** |
| met- | val- | **leu-** | **leu-** |
| asn- | NH$_2$ | asn- | ala- |
| thr | | NH$_2$ | glu- |
| | | | lys- |
| | | | gly- |
| | | | lys- |
| | | | lys- |
| | | | asn- |
| | | | asp- |
| | | | trp- |
| | | | lys- |
| | | | his- |
| | | | asn- |
| | | | ile- |
| | | | thr- |
| | | | gln |

widespread effects such as relaxation of smooth muscles and stimulation of lipolysis and glycogenolysis. The fact that the actions of
VIP on the gut are a minor part of its overall activity and that it is not
released in response to food intake has given rise to the suggestion
that VIP should be considered a neurohormone rather than a gastrointestinal hormone.

## Gastric inhibitory peptide

Gastric inhibitory peptide (GIP) is secreted by K cells located in the
duodenum, jejunum and ileum. The greatest density of K cells is in
the jejunum. GIP contains 43 amino acids, of which seven are
homologous with secretin. The release of GIP is stimulated by the
presence of glucose and triglycerides in the intestinal tract. The
predominant effect of GIP is the stimulation of insulin secretion
by the $\beta$ cells of the islets of Langerhans, although glucose must
also be present for this effect to occur as elevation of GIP concentrations by the administration of triglycerides alone has no effect on
insulin secretion. GIP also inhibits the secretion of gastrin and gastric acid, and it stimulates the release of digestive juices from the
mucosal glands of the duodenum and ileum.

GIP also has actions on the liver, muscles, adipose tissue and the
brain where it potentiates the actions of insulin and acts in a manner contrary to that of glucagon.

## 6.3.3   The peptide family of gastrointestinal hormones

This is a heterogeneous group of peptide hormones that have been
shown to be secreted in the gut, and usually the brain, and which
have some effect on gastrointestinal function. In many cases the
physiological significance of the effects of these hormones on the
gut is unknown.

## Somatostatin

Somatostatin is best known as a hypothalamic hormone with effects
on the secretion of growth hormone (Chapter 12). More recently
somatostatin has been identified in the D ($\delta$) cells of the islets of
Langerhans, and, like VIP, in the D cells of the gastric and intestinal
mucosa. The somatostatin secreted by the cells of the pancreas and
gastric mucosa contains 14 amino acids while that of the intestinal
mucosa is made up of 28 amino acids. Gastrointestinal somatostatin
is secreted in response to the presence of amino acids or glucose in
the gut and in response to stimulation by CCK. Its actions in the gut
are to decrease gastric emptying time and to inhibit the secretion of
gastrin, gastric acid, insulin, glucagon, pancreatic polypeptide and
pancreatic enzymes.

## Motilin

Motilin is secreted by cells throughout the gastrointestinal tract in response to an increase in alkalinity of the duodenum. Its action is to cause slowly travelling contractions along the complete length of the gut although it also causes an increase in the secretion of gastric acid and contraction of the gallbladder. Motilin is also found widely distributed throughout the brain.

## Substance P

The 11 amino acid peptide substance P is widely distributed throughout the gut and the brain; it was the first peptide to be identified in both organs. Within the gut, substance P causes secretion of saliva, increased gut motility and secretion of pancreatic enzymes. It is unlikely that any substance P of gastrointestinal origin enters the systemic circulation.

Several other peptides which may play a role in the control of gastrointestinal activity have been identified in the gut. Gastric acid secretion is decreased by neurotensin, which also inhibits peristalsis, while the secretory activity of the pancreas is inhibited by peptide YY and pancreatic polypeptide, which also causes relaxation of the gallbladder. Enkephalins similarly reduce pancreatic secretion, but also increase gastric acid secretion and cause contraction of the gut, which leads to disruption of peristalsis. Peptide YY, pancreatic polypeptide and enkephalins are all secreted by cells of the gastrointestinal tract following ingestion of food. Both neurotensin and the enkephalins have been shown to play roles as neurotransmitters or neuromodulators.

# 6.4 Co-ordination of gastrointestinal hormone action

As the food enters the stomach there is secretion of gastrin which not only initiates secretion of gastric acid but also stimulates gastric and intestinal motility and induces relaxation of the pyloric sphincter. In readiness for the passage of food into the duodenum, gastrin causes secretion of pancreatic digestive juices, insulin and glucagon. The contraction of the gallbladder which is stimulated by gastrin results in the mixture of bile salts with the gut contents; this not only enables emulsification of fats within the gut contents but is also the route by which products of metabolism are excreted by the liver. Gastrin also promotes the growth of gastric and intestinal mucosa which facilitates the repair of any damage that may have occurred to these cells following exposure to gastric acid and digestive enzymes (see Figure 6.1).

By the time the food enters the duodenum an alkaline environment has been assured by the prior secretion of pancreatic juices. The passage of food into the duodenum results in the release of CCK and secretin which ensure the continued secretion of bile salts and

Figure 6.1

**An illustration of the role of the gastrointestinal hormones in the co-ordination of gastrointestinal tract function and activity**

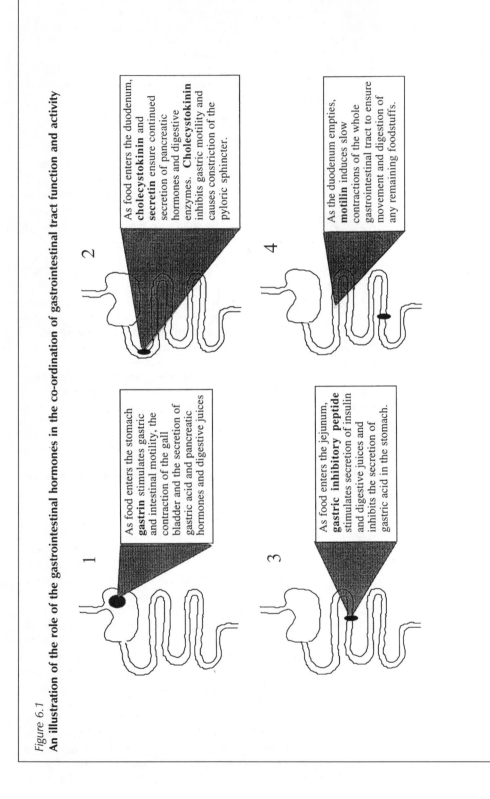

1

As food enters the stomach **gastrin** stimulates gastric and intestinal motility, the contraction of the gall bladder and the secretion of gastric acid and pancreatic hormones and digestive juices

2

As food enters the duodenum, **cholecystokinin** and **secretin** ensure continued secretion of pancreatic hormones and digestive enzymes. **Cholecystokinin** inhibits gastric motility and causes constriction of the pyloric sphincter.

3

As food enters the jejunum, **gastric inhibitory peptide** stimulates secretion of insulin and digestive juices and inhibits the secretion of gastric acid in the stomach.

4

As the duodenum empties, **motilin** induces slow contractions of the whole gastrointestinal tract to ensure movement and digestion of any remaining foodstuffs.

pancreatic enzymes into the gut lumen, and the further secretion of insulin and glucagon. CCK also inhibits the motility of the now empty stomach and causes contraction of the pyloric sphincter to prevent retrograde passage of gut contents.

As the partially digested food passes out of the duodenum and into the jejunum and ileum the secretion of digestive juices by the intestinal mucosa is stimulated by GIP, as is the secretion of insulin which promotes storage or utilization of the absorbed foodstuffs. In a manner similar to that of CCK, further secretion of gastric acid is also inhibited by GIP. The increased amounts of fully digested food-stuffs within the gut (amino acids, glucose and triglycerides) means that maintained secretion of digestive juices is now unnecessary. The presence of such products of digestion thus causes secretion of somatostatin which inhibits release of all digestive juices.

Finally motilin is secreted as the duodenum empties and the pH returns to 4.5. Motilin causes slow peristaltic contractions of the whole gastrointestinal tract to ensure movement and digestion of any residual, undigested foodstuffs. This slow activity continues until the next arrival of food at the stomach, although, in many cases, expectation of the arrival of food, for example as meal times approach, is sufficient to cause the secretion of many of the gut hormones, the initiation of gut contractions and the secretion of digestive juices.

# 6.5 Disturbances of gut hormone function

There are several relatively rare conditions that are associated with abnormalities of gastrointestinal hormone secretion. Most of the conditions involve excessive secretion of the hormone concerned by tumours involving those cell types normally responsible for its secretion. These tumours, however, usually develop not within the intestinal mucosa but either within the pancreas or, occasionally, at sites distant from the gut such as the lungs. There are no major clinical conditions associated with deficiencies of any of the gut hormones with the exceptions of CCK and gastrin.

### 6.5.1 *Cholecystokinin deficiency*

In addition to its actions in the gut, cholecystokinin is also believed to play a role in the brain where it appears to be involved in the sensation of satiety. Some patients suffering from bulimia nervosa, a condition characterized by episodes of excessive, 'binge', eating have been shown to have abnormally low circulating concentrations of CCK. Although these low plasma concentrations reflect decreased secretion by the I cells of the gut, a parallel deficiency may also occur in the brain. It is therefore possible that, in some cases, bulimia nervosa is associated with an impaired secretion of CCK following food intake, and that this results in a lack of satiety, culminating in continued food intake.

In a similar manner, it has been suggested that the reduction of appetite that occurs with aging may be associated with an increased secretion of, or sensitivity to, CCK.

### 6.5.2 Abnormal gastrin secretion

Excess secretion of gastrin results in increased production of gastric acid. In this condition, Zollinger–Ellison syndrome, there may be gastric ulceration and diarrhoea, the latter being caused by the denaturation of the normal digestive enzymes by the hyperacidity. Because gastrin secretion is normally under negative feedback, dependent upon the pH of the stomach contents, attempts to treat Zollinger–Ellison syndrome by the use of antacids or drugs which reduce gastric acid secretion (see later) result in further raised secretion of gastrin. In most cases the gastrin-secreting tumour, 'gastroma', is situated within the pancreas, thus the condition can be treated by surgical removal of the tumour.

In another condition, achalasia, the muscles of the lower oesophageal sphincter are abnormally sensitive to gastrin and are thus maintained in a constricted state. The result of this is that food is unable to pass from the oesophagus to the stomach and therefore accumulates in the oesophagus. The converse condition can occur in cases of gastrin deficiency where the sphincter fails to close and thus allows retrograde passage of the acid stomach contents into the oesophagus resulting in reflux oesophagitis.

### 6.5.3 Vasoactive intestinal peptide excess

Some tumours of the pancreas contain VIP secreting cells, and are thus labelled VIPomas. In this condition, sometimes called pancreatic cholera or Verner–Morrison syndrome, there is inhibition of gastric acid secretion, relaxation of the gastrointestinal smooth muscle and excessive secretion by the pancreas. The major symptom is that of excessive, potentially fatal, watery diarrhoea induced by the gut relaxation coupled with failure of the actions of the digestive enzymes. This failure is brought about by the high alkalinity of the gut as a consequence of the reduced gastric acid secretion and the enhanced bicarbonate secretion by the pancreas. Loss of the bicarbonate ions ultimately results in acidosis and the cardiovascular effects of VIP mean that hypotension is an additional symptom of this condition. VIPomas may respond to treatment with certain anticancer drugs.

### 6.5.4 Somatostatinoma

Intestinal somatostatinomas may cause intestinal obstruction but there are rarely symptoms of somatostatin excess. Pancreatic somatostatinomas, conversely, lead to increased circulating concentrations of the hormone accompanied by a range of gastrointestinal symptoms. The predominant symptom of this condition is mild diabetes mellitus, caused by a selective inhibition of insulin secretion, which is accompanied by delayed gastric emptying and inhibition of gallbladder motility which results in gallstones.

## 6.6 Pharmacological manipulation of gut hormones

Many drugs which act on the gastrointestinal tract are used to treat conditions such as constipation, diarrhoea and gastric or duodenal ulceration; however only those drugs which alter the actions of the gastrointestinal hormones will be discussed here.

Zollinger–Ellison syndrome is characterized by gastric ulceration caused by excessive secretion of gastrin. This condition can be relieved by the use of drugs which reduce the secretion of gastric acid, although they do not treat the underlying condition. The most commonly used antiulcer agents are the histamine ($H_2$) receptor antagonists. These drugs act by preventing the actions of histamine which is the common pathway in the stimulation of gastric acid secretion by gastrin, acetylcholine and by histamine itself (see Figure 6.2). $H_2$ receptor antagonists such as cimetidine and ranitidine increase the incidence of ulcer healing, improve the rate at which healing occurs and prevent the reoccurrence of ulceration. Similar effects can be induced by the use of another group of drugs, the proton pump inhibitors (see Box 6.1). These drugs act by inhibiting the $H^+ K^+$ ATPase of the parietal cells which is responsible for the secretion of the gastric acid. Like the $H_2$ receptor antagonists, proton pump inhibitors such as omeprazole are effective in raising the pH of the stomach and promoting the healing of gastric ulceration.

*Figure 6.2*

**The mechanisms involved in the secretion of gastric acid by oxyntic cells and the effects of the interaction of acetylcholine, histamine and gastrin with their respective receptors**

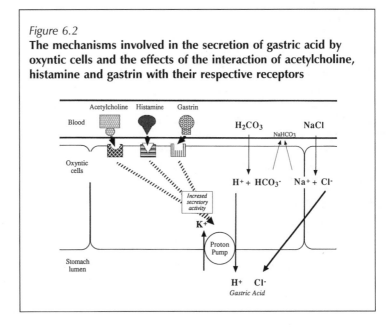

*Box 6.1*  **Drugs used for the treatment of gastrointestinal disorders associated with abnormal secretion of gastrointestinal hormones**

Zollinger–Ellison syndrome is characterized by excessive secretion of gastric acid which results in gastric ulceration. Drugs which inhibit the actions of endogenous histamine or acetylcholine, or which inhibit the proton pump decrease acid secretion and therefore promote healing of the ulceration.

Pentagastrin (an analogue of gastrin and cholecystokinin) may be used to stimulate gastric acid secretion as part of diagnostic investigations of Zollinger–Ellison syndrome.

| *Drug* | *Proprietary name* |
|---|---|
| Histamine (H$_2$) antagonists | |
| Cimetidine | Dyspamet (tablets) |
| Cimetidine | Tagamet (tablets) |
| Famotidine | Pepcid (tablets) |
| Nizatidine | Axid (capsules, injection) |
| Ranitidine | Zantac |
| | |
| Acetylcholine (muscarinic) antagonist | |
| Pirenzepine | Gastrozepin |
| | |
| Proton pump inhibitors | |
| Omeprazole | Losec |
| Lansoprazole | Zoton |

# Summary

- The gastrointestinal tract possesses a diffuse system of cells which secrete hormones responsible for the control of gut motility and the secretion of digestive juices. In many cases the same hormones are also present within the brain where they have distinctly independent roles. This anomaly results from the common embryological origin of the gut and brain cells concerned.

- The gastrointestinal hormones can be divided into three groups: the gastrin–cholecystokinin family which contains gastrin and the cholecystokinins; the secretin family which contains secretin, glucagon, vasoactive intestinal peptide and gastric

inhibitory peptide; and the peptides such as somatostatin, motilin and substance P. Each of the hormones plays a role in the co-ordination of gastric motility, gut peristalsis and digestive enzyme secretion such that the digestion of food and the propulsion of the gut contents along the gastrointestinal tract occurs in a regulated and controlled manner.

- Disorders of secretion of gastrointestinal hormones, or ectopic production by tumours result in the presentation of symptoms such as gastric or duodenal ulcer, diarrhoea or constipation. If left untreated, some of these conditions may prove fatal.

## Selected reading

Binkley, S.A., 1995 Gastrointestinal hormones. In Binkley, S.A., *Endocrinology*, New York: HarperCollins Publishers, 240–257

Hadley, M.E., 1992, Gastrointestinal hormones. In Hadley, M.E., *Endocrinology*, 3rd edn, Englewood Cliffs: Prentice-Hall, 239–269

# 7 The Endocrine Pancreas and Hormonal Control of Carbohydrate Metabolism

## 7.1 Normal carbohydrate metabolism

The normal diet is composed of carbohydrates (sugars), lipids (fats) and proteins. These are absorbed following ingestion and either utilized by the body as a source of energy and in the case of proteins and some lipids, as structural components, or they may be stored. For many of the world's population, that is those who depend on cereals rather than animal fats and proteins for their nutrition, carbohydrates are the major source of energy. During the digestive process these carbohydrates are broken down to produce simple sugars (monosaccharides) which are ultimately converted to glucose before entering the circulation. Glucose is by far the predominant source of energy for the brain and nervous tissue. Within these tissues the glucose is metabolized fully to carbon dioxide and water (glycolysis). Because the brain has very small, if any, stores of carbohydrate, it is dependent on a constant supply of glucose from the blood for its continued function; it ceases to function adequately within two minutes if deprived of its supply of glucose. Despite changes in cerebral activity, the brain's demand for glucose remains relatively constant, thus it is necessary to maintain an adequate supply of blood glucose even though ingestion of carbohydrates is spasmodic. Other groups of cells that require a constant supply of glucose are the red and white blood cells and platelets. These cells are different from nervous tissue however in that they do not metabolize the glucose fully to carbon dioxide and water, but rather only to lactic acid.

Despite the fact that the intake of glucose is spasmodic, being related to meal times, the blood concentrations of glucose are kept relatively constant at approximately 4–8 mmol/l (70–140 mg/100ml). The maintenance of the constant blood concentrations of glucose is achieved by the liver. The liver is able to convert glucose to its storage form glycogen; a process called glycogenesis. Thus after the intake of carbohydrate, some glucose is allowed to enter the circulation immediately, while the remainder is stored in the liver in the form of glycogen. Ninety minutes after the meal, when the blood glucose reservoir starts to become depleted, the liver metabolizes its glycogen stores to produce glucose which is liberated into the bloodstream. The only other organ that is able to break down glycogen and to liberate glucose is the kidney, although it only does this at times of extreme starvation. Another tissue that utilizes glucose as a source of energy is muscle. Unlike the other

> Because the brain has very limited stores of glucose it ceases to function if deprived of a glucose supply for more than two minutes.

Muscles contain enough glycogen to permit 90 minutes of vigorous exercise. The body possesses enough glycogen to overcome 24 hours of fasting.

tissues mentioned, muscle has the ability to produce its own muscle glycogen, thus it has its own carbohydrate store. When required, this glycogen is broken down and metabolized; however, unlike the liver, muscle is unable to release any glucose into the circulation.

Glycogen is the storage form of glucose that is intended for rapid mobilization; under normal circumstances, the body possesses enough glycogen to compensate for approximately 24 hours of fasting. Muscles contain enough glycogen to permit approximately 90 minutes of continued, vigorous exercise. For longer term storage of energy, the body uses fat (see Figure 7.1). Fat can either be obtained from the diet, or it can be synthesized from any carbohydrate that is surplus to immediate requirements (lipogenesis). At rest, both the muscles and the liver utilize fatty acids as their primary source of energy; the muscles only need to use their glycogen during activity. During times of starvation most tissues are able to turn to fats as their predominant source of energy. Metabolism of fats (lipolysis) can be via different routes, one of which results in the liberation of energy and the production of carbon dioxide and water, the other which results in the production of glucose (gluconeogenesis) and ketone bodies. These ketone bodies are then used by some tissues for production of energy. Gluconeogenesis can also utilize amino acids or lactate for the synthesis of glucose. Thus the byproduct of anaerobic carbohydrate metabolism is routinely used as a source for glucose synthesis, as are the products of protein degradation. During times of starvation it is not possible to increase the utilization of proteins as a source of glucose as this would compromise the ability of the individual to survive, thus fats become the major source of energy.

The co-ordination of the processes of glycolysis, glycogenesis, lipogenesis, gluconeogenesis, glycogenolysis and lipolysis is complex; the relative activity of each of the processes is dependent upon the nutritional state of the organism, the activity of the organism and its state of health. Many hormones are able to influence these metabolic processes, but the hormones with greatest influence are those of the endocrine pancreas.

## 7.2 Hormones involved in carbohydrate metabolism

The most important hormones concerned with the control of carbohydrate metabolism are insulin and glucagon; however many other hormones including cortisol, the sex steroids, adrenaline, growth hormone and thyroid hormones all have profound effects. This chapter will concentrate on the actions of insulin and glucagon, although the effects of the other hormones will be discussed where relevant. The synthesis, secretion and mechanism of action of each of the other hormones mentioned is discussed elsewhere in the appropriate chapters.

Both insulin and glucagon are secreted by the pancreas. The pancreas secretes not only these endocrine hormones but also a

Figure 7.1
**Summary of the processes involved in carbohydrate metabolism**

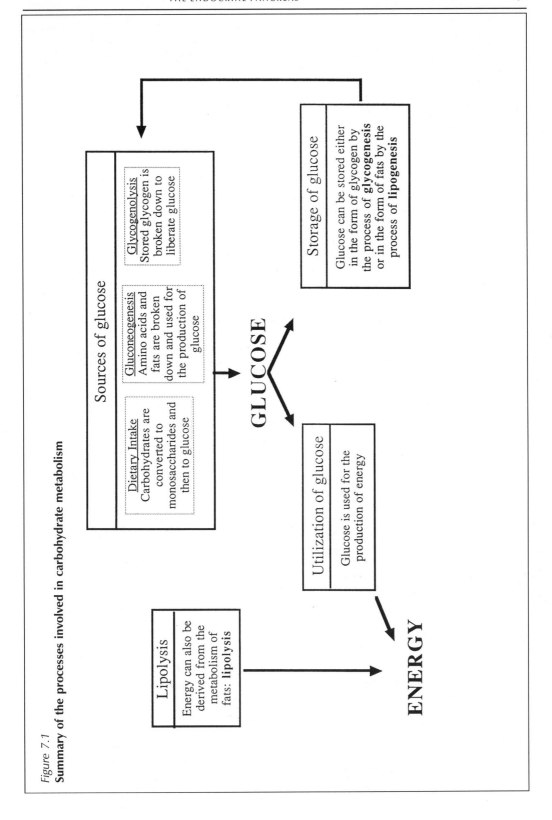

range of exocrine digestive enzymes. Those areas of the pancreas involved with the endocrine aspects are called the islets of Langerhans, clumps of cells which function independently of the rest of the pancreas; less than 2 per cent of the pancreas is concerned with endocrine secretion.

In humans the islets of Langerhans are comprised of three main cell types called $\alpha$, $\beta$ and $\delta$ cells, although they are sometimes refered to as A, B and D cells to avoid confusion with the nomenclature for receptors for adrenaline and the endogenous opioids. The centre of the spherical islets is occupied by $\beta$ cells which are responsible for the synthesis, storage and secretion of insulin; these account for approximately 60 per cent of the total cell number. The $\alpha$ and $\delta$ cells predominate at the periphery of the islet. The $\alpha$ cells are responsible for the synthesis, secretion and storage of glucagon while the $\delta$ cells produce somatostatin. A fourth cell type, the F cells, occur in small numbers and are randomly distributed throughout the islet. These F cells are associated with pancreatic polypeptide.

Within each islet the various cells are very closely associated with each other. Some cells are fused by a gap junction, in which the two cell membranes fuse and form a pore whereby the cytoplasm of the two cells can come into direct contact; other cell–cell junctions involve the membranes of adjacent cells coming into direct contact: tight junctions. This arrangement means that electrical activity spreads rapidly between cells and that there is rapid interchange of ions; the activity of one cell can thus have rapid and profound effects on that of other adjacent cells, not only just of the same cell type, but also of different types.

### 7.2.1  Insulin

Insulin is a polypeptide hormone with 51 amino acids arranged in an $\alpha$ chain of 21 amino acids and a $\beta$ chain of 30 amino acids joined by disulphide bridges. It is synthesized in the $\beta$ cells initially as a large precursor molecule called pre-proinsulin which is then cleaved to form proinsulin which contains 86 amino acids. Proinsulin, the storage form of insulin, is stored within the cell in the form of granules. Immediately prior to secretion proinsulin is cleaved to form insulin and C peptide ('cleavage peptide').

There are several stimuli for insulin secretion but the most important is raised glucose concentration which alters the electrical activity of the $\beta$ cell membranes resulting in the influx of calcium. The raised intracellular calcium induces a contractile process resulting in the movement of the storage granules to the cell membrane, the fusion of the granule with the membrane and the release of the insulin and C peptide into the surrounding fluid.

Insulin secretion occurs in several phases. There is a constant, basal secretion of insulin which occurs even in the absence of any specific stimulus, although it has been shown that this process requires the presence of glucose and there is the release which

The most important stimulus for the secretion of insulin is raised blood glucose concentration.

follows a stimulus. This second form of release is biphasic with an initial release lasting just a few minutes followed, after a decline in insulin secretion, by a more prolonged release which may last for several hours. The initial release probably represents the contents of those granules closest to the $\beta$ cell membrane while the second phase is related to another intracellular pool. It is probable that the contents of the second pool are constantly being replenished by newly synthesized proinsulin. This biphasic release of insulin is even seen when isolated islets of Langerhans cultured *in vitro* are stimulated with glucose.

There are many factors, other than glucose, which can influence insulin secretion. For example administration of glucose orally rather than intravenously results in greater secretion of insulin probably because of the actions of gastrointestinal hormones such as gastrin, secretin and cholecystokinin which are secreted following ingestion of food (see Figure 7.2). Certain amino acids, which are absorbed following the digestion of food, are similarly potent stimulants of insulin secretion. Independent of food intake, the

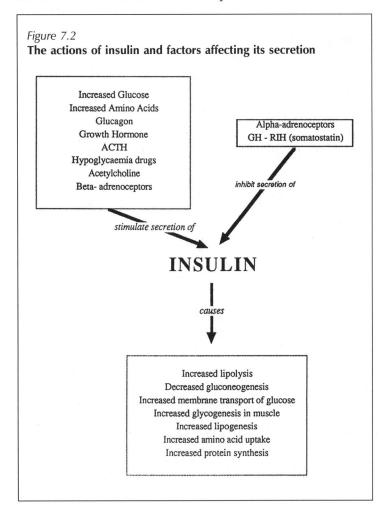

*Figure 7.2*
**The actions of insulin and factors affecting its secretion**

Increased Glucose
Increased Amino Acids
Glucagon
Growth Hormone
ACTH
Hypoglycaemia drugs
Acetylcholine
Beta- adrenoceptors

Alpha-adrenoceptors
GH - RIH (somatostatin)

*inhibit secretion of*

*stimulate secretion of*

**INSULIN**

*causes*

Increased lipolysis
Decreased gluconeogenesis
Increased membrane transport of glucose
Increased glycogenesis in muscle
Increased lipogenesis
Increased amino acid uptake
Increased protein synthesis

autonomic nervous system has an effect. Activity of the sympathetic nervous system inhibits insulin secretion, although this response is actually the net effect of stimulation of $\beta$-adrenoceptors on the $\beta$ cell which potentiates insulin secretion and the stimulation of $\alpha$-adreno-ceptors on the $\beta$ cell which inhibits secretion. At the level of the pancreas the $\alpha$-adrenoceptor effects predominate, but it should be remembered that one result of the stimulation of $\beta$-adrenoceptors by adrenaline is raised blood glucose concentration which would itself precipitate insulin secretion. Stimulation of the parasympathetic nervous system stimulates the secretion of insulin. Other compo-nents of the endocrine system also influence the activity of the islets of Langerhans. Growth hormone and the thyroid hormones increase blood glucose concentrations and therefore indirectly cause secre-tion of insulin, while several of the steroid hormones decrease the effect of insulin at the target tissue; this results in increased plasma glucose and therefore increased insulin secretion. Within the local environment of the pancreas both glucagon and somatostatin may be able to influence insulin secretion by a direct effect on the $\beta$ cells: glucagon in a facilitatory and somatostatin in an inhibitory manner.

### 7.2.2 *Glucagon*

Glucagon contains 29 amino acids and in a manner similar to insu-lin is initially synthesized as pre-proglucagon and, prior to release by exocytosis, is stored in granules within the $\alpha$ cells as progluca-gon. Unlike insulin, there is a high basal secretion of glucagon, and, in general, factors which increase insulin secretion decrease the secretion of glucagon (and vice versa); thus glucagon secretion is potentiated by a decrease in plasma glucose concentration and inhibited by an increase. The mechanism of this effect is unknown. Similarly, a decrease in circulating fatty acids stimulates glucagon secretion while an increase in fatty acids is inhibitory. Stimulation of the sympathetic nervous system increases glucagon secretion (see Figure 7.3).

Paradoxically, some stimuli have the same effects on the se-cretion of both glucagon and insulin. Amino acids stimulate the secretion of glucagon as well as the stimulation of insulin, as does cholecystokinin (see above). Similarly stimulation of the parasym-pathetic nervous system stimulates the release of both hormones. At the local level of the pancreas, both insulin and somatostatin inhibit glucagon secretion.

### 7.2.3 *Somatostatin and pancreatic polypeptide*

The somatostatin produced by the $\delta$ cells is a 14 amino acid peptide which is identical to the somatostatin of the hypothalamus (Chapters 2 and 12). Like insulin and glucagon it is initially synthe-sized as a preprohormone; however immunoreactive somatostatin, rather than a prohormone, is identifiable within the $\delta$ cells. Release of pancreatic somatostatin occurs following ingestion of proteins.

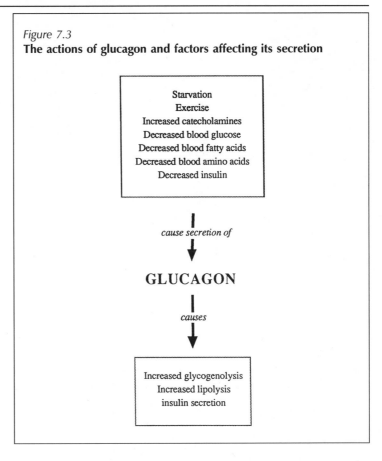

Figure 7.3
**The actions of glucagon and factors affecting its secretion**

Starvation
Exercise
Increased catecholamines
Decreased blood glucose
Decreased blood fatty acids
Decreased blood amino acids
Decreased insulin

*cause secretion of*

**GLUCAGON**

*causes*

Increased glycogenolysis
Increased lipolysis
insulin secretion

Like somatostatin, pancreatic polypeptide, which is made up of 36 amino acids, is secreted following ingestion of proteins. Similar elevations in pancreatic polypeptide occur during hypoglycaemia and following stimulation of the parasympathetic nervous system. The full details of the synthesis, storage and release of somatostatin and pancreatic polypeptide are unknown although it is apparent that somatostatin inhibits the secretion of pancreatic polypeptide and that pancreatic polypeptide inhibits the secretion of somatostatin.

## 7.3 Hormonal influence on carbohydrate metabolism

Many hormones other than insulin and glucagon influence carbohydrate metabolism: they all raise blood glucose concentrations.

As stated earlier, many hormones are able to influence the metabolism of carbohydrates, and these effects are brought about by a variety of mechanisms. With the exception of insulin, the effect of all of these hormones is to increase blood glucose concentrations; insulin lowers blood glucose. The mechanisms of action and the effects of each of the major hormones that influence carbohydrate metabolism will be described in turn.

### 7.3.1  *Insulin*

The overall effect of insulin is to lower blood concentrations of glucose by increasing the uptake of glucose from the blood into tissues such as muscle, by promoting the storage of glucose in the form of glycogen and fat and by decreasing the liberation of glucose by glycogenolysis, gluconeogenesis and lipolysis. Insulin also promotes amino acid uptake and protein synthesis.

Insulin produces its effects by interaction with a membrane bound, tyrosine kinase linked receptor (see Chapter 1). Unlike most tyrosine kinase linked receptors which are single protein chains which span the cell membrane, the insulin receptor is a dimer of two identical membrane spanning chains. The binding of insulin to the extracellular portion of the receptor has several effects. The first of these is the movement of glucose transporters to the cell surface. All cells possess glucose transporters which facilitate the transport of glucose across the cell membrane against a concentration gradient. One of the transporters, $GLUT_4$, is an insulin sensitive transporter that is found in cardiac and skeletal muscle and in adipose tissue. The interaction of insulin with its receptor causes the $GLUT_4$ transporters to move from their intracellular 'storage' sites to the cell membrane where they promote glucose uptake.

Another effect of the binding of insulin to its receptor is the stimulation of the phosphorylating activity of the insulin receptor. Phosphorylation of certain enzymes alters their activity, for example the phosphorylation of glycogen synthase in the liver results in increased synthesis of hepatic glycogen while inhibition of hepatic glucokinase decreases the liberation of glucose by glycogenolysis. There is also evidence that the second messengers cAMP and cGMP may mediate the effects of insulin in some cells. In hepatic cells, insulin decreases the concentration of cAMP which results in a decrease in glycogenolysis and gluconeogenesis. In adipose tissue insulin may also reduce lipolysis by reducing formation of cAMP.

As described in Chapter 1, the binding of a hormone to a tyrosine kinase linked receptor stimulates protein phosphorylation which may alter the activity of certain cell transcription factors thus influencing protein synthesis, cell differentiation and cell replication. Insulin enhances protein synthesis in some cells and increases the active transport of amino acids into cells. When this is considered together with the fact that insulin decreases the utilization of proteins for gluconeogenesis, it can be seen that the presence of insulin increases the availability of proteins for growth and development.

As is the case for other hormone receptors, the binding of insulin to its membrane bound receptor is followed by invagination of a portion of the membrane to produce vesicles containing the hormone–receptor complex. This process facilitates the removal of insulin from its receptor and allows the receptor to be recycled to the membrane; the insulin may then be destroyed by proteolysis. The importance of the receptor internalization process is that it may be involved in receptor down regulation. It has been demonstrated that at times of raised circulating concentrations of insulin, the number of insulin receptors at the cell surface is decreased, and it is possible that this process is important in the aetiology of some forms of diabetes mellitus (see later).

### 7.3.2 *Glucagon*

Glucagon increases blood glucose by enhancing hepatic glycogenolysis and by stimulating transport of amino acids into the liver for gluconeogenesis. Lipolysis within adipose tissue is also stimulated. These effects are brought about following the interaction of glucagon with a specific membrane bound receptor. This receptor is a single protein chain which spans the cell membrane, the stimulation of which results in the synthesis of the intracellular second messengers cAMP or $IP_3$. Most of the actions of glucagon appear to be mediated by cAMP.

### 7.3.3 *Other hormones*

Several other hormones have direct effects on carbohydrate metabolism. An appreciation of these effects is necessary for the understanding of some of the endocrine related disorders of carbohydrate metabolism; however their effect will be discussed in more detail in the appropriate chapters.

Adrenaline (Chapter 8) produces a rapid increase in blood glucose as part of its 'fight or flight' response. Thus, via actions on the $\beta$-adrenoceptor, adrenaline stimulates hepatic glycogenolysis and gluconeogenesis and also stimulates glycogenolysis in skeletal muscle although the latter does not result in liberation of glucose. In adipose tissue, lipolysis is promoted resulting in the mobilization of fatty acids. Via actions on the $\alpha$-adrenoceptor, catecholamines inhibit the release of insulin from the pancreas.

Thyroid hormones (Chapter 4) and growth hormone (Chapter 12) both increase blood glucose. In the case of growth hormone this is a result of enhanced lipolysis in adipose tissue while in the case of the thyroid hormones all of the events of carbohydrate metabolism are accelerated; thus there is increased glycogenolysis, gluconeogenesis and lipolysis as well as enhanced glycogenesis and lipogenesis; the net effect is catabolic and there is thus an increase in blood glucose. Steroid hormones (Chapter 9), most notably cortisol, also increase blood glucose. These hormones stimulate the synthesis of enzymes which stimulate gluconeogenesis and lipolysis, they also reduce the effects of insulin at its receptor possibly as a result of insulin receptor down regulation as a consequence of the prolonged elevation of blood glucose concentration.

Adrenaline causes a rapid increase in blood glucose concentration as part of the 'fight or flight' response.

# 7.4 Disorders in the endocrine control of carbohydrate metabolism

## 7.4.1 Hypoglycaemia

Hypoglycaemia may occur following prolonged fasting, excessive exercise or as a rebound following carbohydrate intake. For example it has been shown that blood glucose may fall below normal levels two to four hours after a carbohydrate meal; the reason for this is that the initial rapid food intake produces insulin secretion which may be slightly in excess of that required to remove the glucose load. Such reactive hypoglycaemia has been shown to occur in 20 per cent of normal subjects without the manifestation of any symptoms. The most common cause of pathological hypoglycaemia is an insulin excess. This may occur as a result of a pancreatic tumour, which may be associated with a multiple endocrine neoplasia (see Chapter 5), as a result of an autoimmune disease in which antibodies are able to stimulate insulin secretion; or it may be due to ectopic secretion of insulin or insulin like growth factor (see Chapter 12) by a large tumour. Another common cause is insulin overdose, or insufficient carbohydrate intake, in people receiving insulin therapy (see later).

The symptoms of hypoglycaemia are generally a consequence of the decreased supply of glucose to the central nervous system. Thus there is lethargy, dizziness, confusion, blurred vision and incoherent speech. Ultimately the condition may result in coma or death. In many cases there are simultaneous symptoms of autonomic nervous system stimulation, these being sweating, tremor and palpitations (see Box 7.1). In the case of insulin overdose or insufficient carbohydrate intake the symptoms are relieved rapidly following administration of glucose. Where the condition is a result of an insulin secreting tumour the only treatment is the surgical removal of that tumour although use of the drug diazoxide which destroys islet tissue may also be effective.

> Hypoglycaemia is signalled by lethargy, confusion, dizziness and slurred speech.

## 7.4.2 Hyperglycaemia

Diabetes mellitus is the name given to a syndrome in which the prominent feature is hyperglycaemia. It was already well described over 2000 years ago, indeed its name is a description of its most obvious symptom: the term diabetes is derived from a Greek word signifying a syphon and mellitus is derived from the Latin for

Box 7.1 **Features of hypoglycaemia**

| Causes of hypoglycaemia | Symptoms of hypoglycaemia | Treatment of hypoglycaemia |
|---|---|---|
| Insufficient carbohydrate intake | Blurred vision | Administration of glucose |
| Excessive dose of insulin | Confusion | |
| Excessive exercise | Sweating | |
| | Shaking | |

Diabetes means syphon; mellitus means honey.

honey. The disease is generally classified into several different groups which have different aetiologies and treatments but for which the symptoms are similar. The first order of classification is into primary and secondary diabetes mellitus. In primary diabetes there is either a lack of insulin or a failure of the target tissues to respond to insulin, while secondary diabetes mellitus may be caused by inappropriate secretion of one of the other hormones that influence carbohydrate metabolism, for example excess secretion of corticosteroids in Cushing's disease (see Chapter 9) or the raised secretion of oestrogens during pregnancy (see Chapter 11).

Primary diabetes mellitus is then further divided into two groups. This secondary division was originally based upon the age of onset of the disorder; thus if the condition was diagnosed before the age of 25 it was termed juvenile onset diabetes mellitus (JOD) and if it was diagnosed after 25 it was termed maturity onset diabetes mellitus (MOD). This classification arose because it was noted that the two conditions could be differentiated in terms of some of the features of symptomatology and the response to treatment (see later). It was subsequently noted that in some cases, patients developed a disorder before the age of 25 which exhibited symptoms more akin to those of MOD. This gave rise to a further classification of maturity onset diabetes mellitus of the young (MODY). Furthermore some patients showed no symptoms of diabetes mellitus until the age of 30 when they developed a disorder characteristic of JOD. The nomenclature was therefore changed to type I diabetes mellitus for JOD and type II diabetes for MOD and MODY. More recently the terms insulin dependent diabetes mellitus (IDDM) and non-insulin dependent diabetes mellitus (NIDDM) have replaced type I and type II respectively.

There are two different forms of diabetes mellitus; one usually onsets before the age of 25, the other after.

Insulin dependent diabetes mellitus, or type I, is a direct consequence of the failure of the pancreas to secrete adequate quantities of insulin. The onset of this disorder commonly ocurs at or before the time of puberty. One possible cause of this disorder is an autoimmune disease in which individuals produce antibodies against their own pancreatic tissue; the appearance of such antibodies is often noted following infection with viruses such as the mumps virus and cocksackie virus B4, triggers known to be associated with an increased incidence of IDDM. Similar antibodies, however, are found in a significant proportion of the unaffected (asymptomatic) population suggesting that other factors may be involved in the development of IDDM. One possibility is that there is a genetic predisposition to IDDM as there is a 30 per cent concordance rate in the incidence of IDDM between identical twins and 5–8 per cent prevalence in first degree relatives of sufferers. It has been suggested that one possible genetic factor is the absence of the gene responsible for pancreatic regeneration. It is believed that the islet cells of normal individuals are constantly being broken down and regenerated, and that there is a specific gene responsible for this ability to regenerate the $\beta$ cells; deficiency of this gene would therefore ultimately result in the development of IDDM. It is unlikely, however, that this is the only cause of IDDM. Approximately 1 per cent of the population suffer from IDDM although it is particularly prevalent (up to 15 per cent) in South Pacific islanders and native Americans. Diabetes mellitus may also occur alongside other endocrine deficiencies such as Addison's disease (Chapter 9), Graves' disease (Chapter 4) or hypogonadism (Chapters 10 and 11).

Approximately 1 per cent of the population suffers from insulin dependent diabetes mellitus.

The aetiology of NIDDM is very different. In these patients there may be circulating concentrations of insulin which are above normal values; the patients however exhibit the symptom of insulin deficiency. In these patients the pancreas is functioning but the target tissues fail to repond to the insulin. In some cases this is due to the presence of anti-insulin antibodies which inactivate the hormone, but in the majority of cases it is a result of insulin receptor down regulation. NIDDM is prevalent among obese individuals, where the persistently high blood glucose results in internalization, and therefore inactivation, of the insulin receptor. Simultaneously, it has been suggested that the excessive lipid stores within the adipose tissue in some way decreases the insulin receptor activity. Approximately 10 per cent of the population in the Western world develops some degree of NIDDM during later life.

Ten per cent of the population of the developed world suffers from non-insulin dependent diabetes mellitus.

## Symptoms and consequences of diabetes mellitus

The symptoms and consequences of IDDM and NIDDM are generally similar and will be considered together although there are differences in the general appearance of the sufferer. The onset of IDDM is usually rapid and generally occurs in young people who are of slight build. NIDDM occurs in overweight individuals, often in the fifth or sixth decade of life. The appearance of the symptoms is gradual and there is usually evidence of raised blood glucose for several years before diagnosis.

Hyperglycaemia itself is symptomless other than the polyuria and polydipsia. Polyuria (output of increased volume of urine) occurs because the raised blood glucose concentration exceeds the renal threshold for glucose reabsorption. Under normal circumstances all of the glucose that is filtered at the glomerulus is reabsorbed by an active transport system in the proximal tubule of the nephron. This transport system is saturable; thus if the blood glucose concentration exceeds a value of about 10 mmol/l, glucose will be excreted in the urine. Because of the osmotic effect of the glucose, as it is excreted it is accompanied by an increased volume of water. This therefore accounts for the polyuria. Dehydration occurs as a consequence of the increased flow of urine, and this results in the sensation of thirst and the increased fluid intake (polydipsia). One further consequence of the excretion of glucose in the urine is the increased incidence of genito-urinary infections because of the favourable conditions for bacterial growth. Wounds may also become infected more easily because of the high glucose content of the blood. These symptoms themselves are relatively minor although they are indicative of a more serious underlying condition.

The lack of insulin, or lack of response to insulin, means that tissues are unable to take up or to utilizse glucose. In the case of muscle this means that all of the energy must be obtained by metabolism of fat; this is of particular importance in IDDM where there is an absolute deficiency of insulin. The inability to use muscle glycogen as a source of energy means that there is a loss of muscle strength and a feeling of lethargy, while the utilization of the alternative route of metabolism results in the production of lactic acid and ketone bodies which culminates in ketonaemia, the symptoms of which are nausea and vomiting, and acidosis; some of the excess ketone bodies are excreted in the expired air giving a characteristic smell to the breath. The utilization of muscle protein as a substrate for gluconeogenesis ultimately results in muscle wasting and weight loss. These symptoms are less serious in NIDDM due to the presence of some residual insulin activity.

The starvation of cells of glucose, coupled with the disturbance of the acid–base balance and the ketonaemia leads to irreversible damage to several delicate tissues. One of the first tissues to become damaged is the endothelium of the blood vessels. This damage renders the blood vessels incompetent and results in peripheral vascular disease. Damage to the microvessels of the retina may result in blindness, while damage to the vessels supplying the fingers and toes may cause gangrene. A common consequence of long-term, uncontrolled diabetes mellitus is the necessity to amputate parts of the lower limb. Nerve cells may also become damaged; if this occurs in the autonomic nervous system there may be postural hypotension, disorders of gut motility, loss of bladder control and impotence. Damage to sensory fibres leads to loss of sensation, particularly in the hands and feet. A further consequence of the abnormal route of carbohydrate metabolism in diabetes mellitus is that cataracts develop more rapidly in diabetics than in the normal population.

Sadly, the overall picture of a patient with chronically uncontrolled diabetes mellitus is one of someone with failing eyesight (due to retinopathy and cataracts), who is breathless and easily fatigued (due to abnormal carbohydrate metabolism and acidosis), frequently complains of cold hands and feet (peripheral vascular disease) and may have severe lesions, such as foot ulcers or gangrene, of their feet which are unrecognized because of the loss of sensation (see Box 7.2). Most of these symptoms are irreversible, however their onset can be delayed considerably, possibly indefinitely, by careful control of the underlying diabetes mellitus.

Long-term, uncontrolled diabetes mellitus results in blindness and gangrene.

## Treatment of diabetes mellitus

As the name suggests, the treatment of IDDM is dependent upon insulin replacement. To date there is no other form of medication that is able to overcome the pancreatic deficit. Because of the role of

*Box 7.2*  **Features of diabetes mellitus**

*Early signs of diabetes mellitus*
Lethargy
Weight loss
Increased appetite
Polydipsia
Polyuria

Biochemical tests would reveal hyperglycaemia and glycosuria

*If left untreated, complications of diabetes mellitus are:*
Blindness
Cataracts
Peripheral vascular disease, possibly causing gangrene
Peripheral neuropathy, possibly causing loss of sensation, incontinence and impotence
Cardiovascular disease
IDDM may rapidly result in death

insulin in the control of carbohydrate metabolism, it is also necessary to exercise some control over dietary intake (see later). In NIDDM it is sometimes possible to reduce the severity of the diabetes mellitus by dietary manipulation alone. Where this is unsuccessful aditional therapy with an oral hypoglycaemic drug, or with insulin, is required. Each of these forms of therapy will be discussed in turn.

Insulin was first isolated from the pancreas by Banting and Best in 1923. They were not the first to note the hypoglycaemic effects of a pancreatic extract, but earlier workers had been unable to extract the active principle successfully because of the digestive enzymes released during the maceration of the non-endocrine pancreas.

The earliest form of insulin to be used therapeutically was obtained from pancreases from cows or pigs. These bovine and porcine insulins are active in humans despite the fact that they differ from human insulin; in one or three amino acids respectively. Modern therapeutics has switched to replacement therapy using human insulin; this has the advantage that it is not recognized as a foreign protein and therefore does not precipitate antibody formation. This human insulin is made either by the enzymatic conversion of porcine insulin, or by biotechnological procedures using either *Escherichia coli* or yeast (*Saccharomyces cerevisiae*).

Because insulin is a peptide it cannot be administered orally; the most common form of administration is therefore by injection, the usual sites of injection being the upper arms, thighs, lower abdomen or the buttocks. Once injected the insulin undergoes distribution within the blood to the target tissues before it is degraded, predominantly by the liver. Following subcutaneous or intramuscular administration insulin can be detected in the blood for up to three hours. Different formulations of insulin have been developed to vary the duration of action of the injected insulin (see Figure 7.4). For example the injection of a simple solution of insulin may be

## Figure 7.4
**Available formulations of insulin for the treatment of diabetes mellitus (reproduced from the *Monthly Index of Medical Specialities (MIMS)*, in which details of available preparations are updated monthly, with permission of the publisher)**

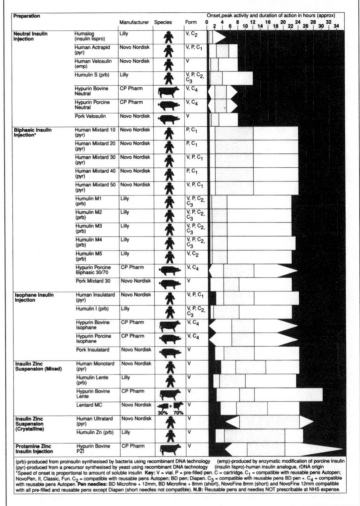

sufficient to lower blood glucose for a period of one to four hours; therefore if such treatment were to be used in the treatment of IDDM it would necessitate up to six injections per day. This necessity has been overcome by the development of long-acting insulins by the addition of either zinc or the protein protamine to the insulin solution. These long-acting insulins have durations of action of up to 24 hours. In practice most diabetics are treated with a combination of a short-acting and a long-acting insulin.

By careful monitoring of blood glucose concentrations and diligent adjustment of the dose of insulin it is possible to maintain glucose concentrations within the normal range and thus to prevent the onset of the complications of uncontrolled diabetes mellitus. Such a treatment regime, however, would be very time consuming and invasive. The usual form of treatment, therefore, is to inject fixed doses of insulin at set times of day and to keep to a relatively controlled diet. In this way the amount of insulin required to remove the ingested load of carbohydrate from the bloodstream can be predicted. The efficacy of this form of treatment can then be monitored by occasional determinations of urine or blood glucose. In most cases the patient then has the freedom to adjust the dose of insulin or the diet according to the results of the glucose monitoring. For many people this form of therapy is very effective, but problems can occur with patient behaviour (see Box 7.3).

In IDDM, food intake must be matched to the administered dose of insulin. Insufficient intake of carbohydrate may result in hypoglycaemia, the symptoms of which have been described earlier, while excessive carbohydrate intake would result in hyperglycaemia. The dose of insulin and the diet must also be matched to the predicted physical activity to be undertaken, thus a sedentary office worker would require a lower food intake and insulin dose than a manual labourer.

Dietary control is also very important in NIDDM. As described above, one of the major causes of NIDDM is obesity and persistently raised blood glucose. By restricting carbohydrate intake it has been

---

*Box 7.3* **The use of insulin in the treatment of diabetes mellitus**

As outlined above, the early stages of diabetes mellitus are relatively symptom free other than polyuria and polydipsia, therefore there are few early symptoms of poorly controlled diabetes mellitus; patients thus feel no ill effects if they fail to take to their prescribed treatment regime. Because of the inconvenience and discomfort of regular insulin injection, coupled with the necessity of carrying around syringes and medication, many diabetics, especially younger ones, are tempted to treat themselves only erratically. The threat of possible severe complications of uncontrolled diabetes mellitus in many years time is insufficient to ensure adequate control of blood glucose. Prior to attendance at routine diabetic clinics the patients may attempt to regain control of their blood glucose, but previous excursions from normal glucose concentrations can now be detected. If blood glucose remains high for a long period of time, the glycosylation of some blood components can occur. This fact has been used in the development of assays for fructosamine and glycosylated haemoglobin

which have been shown to be reliable in the monitoring of glycaemic control over the previous 6–12 weeks.

Other developments have been made in order to increase patient compliance among insulin users. The first of these is constant subcutaneous insulin infusion (CSII). In this technique insulin is delivered via an indwelling subcutaneous cannula by a slow infusion pump. The delivery of insulin is such that there is a constant baseline infusion onto which is superimposed a boost at the time of food intake. Using this form of administration the blood levels of insulin closely resemble those seen in normal, non-diabetic individuals, and very good glycaemic control is achieved. Another development has been the insulin pen. This is a spring-loaded injection device prefilled with a set dose of insulin. These pens have been shown to be quick and effective for the injection of insulin and have the advantage that they only require the use of one hand. There have also been several attempts to develop an orally active insulin, but as yet none have been successful.

shown that a return of normal insulin function can accompany a return to normal body weight. Dietary control is the first line of treatment in NIDDM, but failure to respond to this form of treatment may necessitate the use of oral hypoglycaemic drugs.

Oral hypoglycaemic agents are drugs which, when taken orally, confer some control over carbohydrate metabolism. Their use is limited by the fact that they appear to require a functioning pancreas in order to produce their effect; thus they can only be used in the treatment of NIDDM, not IDDM. In the treatment of NIDDM these drugs have their greatest effect on basal blood glucose concentration, they have little effect on the raised blood glucose that occurs after food intake, thus their use is usually coupled with dietary control.

The most commonly used group of oral hypoglycaemic agents are the sulphonylurea drugs such as chlorpropamide, glibenclamide, tolbutamide and glipizide (see Box 7.4). These drugs bind to the surface of the $\beta$ cells of the islets of Langerhans to stimulate insulin secretion. They appear to do this by interacting with specific membrane bound receptors which results in a decrease in potassium efflux, partial depolarization of the cell membrane and influx of calcium ions. It is this calcium influx which causes the secretion of insulin (see Figure 7.5). These drugs also sensitize the $\beta$ cells to stimuli that cause insulin secretion possibly by raising intracellular levels of the second messenger cAMP as a consequence of inhibition of phosphodiesterase. In high concentrations, sulphonylureas increase insulin mediated tissue uptake of glucose and increase insulin receptor density, although these actions are probably irrelevant at therapeutic doses. These drugs are effective in the treatment of many cases of NIDDM, but they are not without side-effects, one of which is that they tend to produce an increase in weight, and this is in a population being treated for a disorder which may be a consequence of obesity.

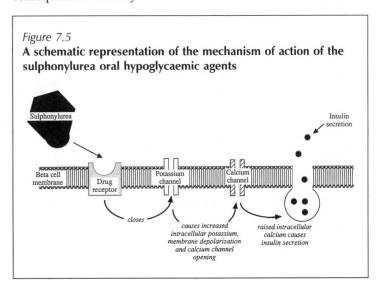

*Figure 7.5*
**A schematic representation of the mechanism of action of the sulphonylurea oral hypoglycaemic agents**

Box 7.4    **The use of oral hypoglycaemic agents in the treatment of diabetes mellitus**

There are two classes of oral hypoglycaemic agents which both act to decreases blood glucose concentrations, although by different mechanisms. Both classes, however, require some residual pancreatic secretion of insulin in order to produce their effects; they cannot, therefore, be used in the treatment of insulin dependent diabetes mellitus.

In the treatment of non-insulin dependent diabetes mellitus the first action is usually to restrict the intake of carbohydrates and other sources of energy; in many cases the weight loss that is induced by this action is sufficient to promote normal control of glucose metabolism. Only if dietary restrictions fail to control the hyperglycaemia is the use of oral hypoglycaemic agents considered.

| Drug | Proprietary name |
|---|---|
| *Sulphonylureas* | |
| Chlorpropamide | Diabinese (tablets) |
| Glibenclamide | Daonil (tablets) |
| Glibenclamide | Euglucon (tablets) |
| Glicazide | Diamicron (tablets) |
| Glipizide | Glibenese (tablets) |
| Glipizide | Minodiab (tablets) |
| Gliquidone | Glurenorm (tablets) |
| Tolazamide | Tolanase (tablets) |
| Tolbutamide | Rastinon (tablets) |
| *Biguanides* | |
| Metformin | Glucophage (tablets) |

Another group of oral hypoglycaemic agents is the biguanide drugs, of which the only one available for clinical use is metformin. This drug acts by enhancing peripheral uptake of glucose and by reducing gluconeogenesis. It is an effective drug but its usefulness is limited by the fact that it sometimes causes profound lactic acidaemia.

## Summary

- For many people the predominant source of energy is carbohydrate which is ultimately converted to glucose. The intake of carbohydrate, which occurs at meal times, is often sporadic, however the body requires the concentration of circulating glucose to be maintained within close limits in order to maintain adequate function. This control is achieved by the conversion of glucose to storage forms, glycogen and fat, during times of relative plenty, and the utilization of these storage forms during fasting (for example 90 minutes after a meal). This balance of utilization, storage and liberation of glucose is achieved by the interplay of several hormones, most notably insulin and glucagon.

- At times of raised plasma glucose concentrations, insulin is secreted from the islets of Langerhans. The effect of the insulin is to promote the uptake and utilization of glucose by cells and to enhance the production of storage of glucose in the form of glycogen and fat. As a result of the action of insulin, plasma glucose concentration falls, and therefore insulin secretion declines. At times of decreased plasma glucose, glucagon is secreted from the islets of

Langerhans. Glucagon initiates the mobilization of glucose stores by enhancing the breakdown of glycogen and fats. Another important action of glucagon is the stimulation of insulin secretion, which allows utilization of the liberated glucose.

- A deficiency of glucagon or an excess of insulin results in hypoglycaemia, the symptoms of which reflect the decreased activity of the brain as a consequence of the reduced supply of energy: lethargy, dizziness and confusion. Excess glucagon or a deficiency of insulin results in raised plasma glucose and decreased glucose utilization. This gives rise to diabetes mellitus, which can affect up to 10 per cent of the population. Diabetes mellitus can be caused by a failure of insulin secretion (insulin dependent diabetes mellitus), a failure of the tissues to respond to insulin (non-insulin dependent diabetes mellitus) or a disorder of some of the other hormones known to influence carbohydrate metabolism (secondary diabetes mellitus). Long-term diabetes mellitus may result in blindness, neuropathy and gangrene, which may necessitate amputation of a limb, although these consequences may be prevented by adequate control of carbohydrate metabolism by use of either insulin replacement therapy, dietary control or oral hypoglycaemic agents.

## Selected reading

Aitman. T.J. and Todd, J.A., 1995, Molecular genetics of diabetes mellitus. In Thakker, R.V. (ed.), *Baillière's Clinical Endocrinology and Metabolism, International Practice and Research. Vol. 9, No. 3, Genetic and Molecular Biological Aspects of Endocrine Disease*, London: Baillière Tindall, 631–656

Jones, D.B. and Gill, G.V., 1997, Insulin-dependent diabetes: an overview. In Pickup, J.C. and Williams, G. (eds) ,*Textbook of Diabetes*, 2nd edn, Oxford: Blackwell Science, 12.1–12.13

Jones, D.B. and Gill, G.V., 1997, Non-insulin-dependent diabetes: an overview. In Pickup, J.C. and Williams, G. (eds), *Textbook of Diabetes*, 2nd edn, Oxford: Blackwell Science, 17.1–17.13

Lawrence, J.C. 1994, Insulin and oral hypoglycemic agents. In Brody, T.M., Larner, J. and Neu, H.C. (eds), *Human Pharmacology: Molecular to Clinical*, 2nd edn, St Louis: Mosby, 501–514

Laycock, J. and Wise, P., 1996, The pancreas (1): control of metabolism, diabetes, and hypoglycaemia. In Laycock, J. and Wise, P., *Essential Endocrinology*, 3rd edn, Oxford: Oxford University Press, 274–314

# 8 The Adrenal Medulla

## 8.1 Introduction

The adrenal glands lie in the abdominal cavity, directly above the kidneys, hence the name suprarenal gland which is occasionally used in the USA. The glands are each divided into two distinct parts, the inner adrenal medulla (10 per cent) and the outer adrenal cortex (90 per cent). These two portions function independently of each other, although their actions are related.

Embryologically the outer adrenal cortex is derived from the mesoderm while the adrenal medulla is derived from the neural crest of the ectoderm, thus the cells of the adrenal medulla are closely related to cells of the nervous system. These medullary cells are usually referred to as chromaffin cells because of their affinity for histological stains. Although the two portions of the adrenal glands are independent, the activity of one is unavoidably influenced by the activity of the other because of the anatomy of their nerve and blood supply. The nerve supply of the adrenal medulla is derived from the greater splanchnic nerve and is via a preganglionic sympathetic neurone. There is no postganglionic portion to this innervation; that role is played by the adrenal medulla itself, thus this arrangement is unusual for the sympthetic nervous system as the neuroeffector junction (i.e. the nerve–gland junction) utilizes acetylcholine as its transmitter rather than the usual sympathetic neurotransmitter noradrenaline. There is also some evidence that a minor portion of this nerve supply is used to innervate the adrenal cortex, although some of the cells of the adrenal cortex are innervated by neurones originating in the adrenal medulla. The blood supply to the adrenal gland is provided by a variety of small vessels, some of which are branches of the renal artery and the aorta. Within the adrenal gland many of these vessels anastomose so that there is a very rich capilliary network within the outer layers of the cortex. The inner layers of the cortex and the adrenal medulla receive a partially deoxygenated blood supply that has previously passed through other areas of the gland and therefore contains the endocrine secretions of these areas. Blood is drained from the right adrenal gland into the inferior vena cava, while that of the left adrenal gland enters either the renal vein or the inferior phrenic vein.

## 8.2 Synthesis and secretion of adrenal medullary hormones

Several peptide hormones have been identified within the chromaffin cells, for example enkephalins and vasopressin, but by far the most important hormones secreted by the adrenal medulla are the catecholamines adrenaline and noradrenaline, respectively termed epinephrine and norepinephrine in the USA. Approximately 80 per cent of the chromaffin cells are of the type which synthesize adrenaline. The route of synthesis of these catecholamines by the adrenal gland is identical to that found in the postganglionic sympathetic nerve fibres and the brain (see Figure 8.1). The starting point for the synthesis is the amino acid tyrosine which is usually obtained from the diet but which may be formed from phenylalanine in the liver. The initial stage of synthesis is the conversion of tyrosine to dihydroxyphenylalanine, more commonly called dopa, by the enzyme tyrosine hydroxylase. The dopa is then rapidly converted to dopamine by dopa decarboxylase. Dopamine is a neurotransmitter in its own right within the brain, but within the adrenal medulla the activity of the enzyme dopamine-$\beta$-hydroxylase is such that it is immediately converted to noradrenaline, thus it is not a secretion of the adrenal medulla. In a small number of chromaffin cells the pathway ceases with the production of noradrenaline, but in the majority of the cells the noradrenaline is methylated to adrenaline by the action of phenylethanolamine-N-methyltransferase.

> The adrenaline and noradrenaline of the adrenal medulla are synthesized by the same route as that found in the brain.

Once synthesized the adrenaline and noradrenaline are stored in the chromaffin cells within granules. Secretion of these hormones is stimulated by the neurotransmitter acetylcholine which is released form the preganglionic sympathetic neurones which innervate the chromaffin cells (see above); the acetylcholine causes exocytosis of the granules. The adrenal medulla can therefore be considered as a modified form of a postganglionic sympathetic nerve fibre, differing from the nerve fibre only in that the predominant secretion is adrenaline rather than noradrenaline and that the transmitter/hormone is released into the blood stream rather than into a neuro-effector junction. It therefore follows that the secretion of adrenaline and noradrenaline by the adrenal medulla is increased by any form of general sympathetic nervous system stimulation.

> The adrenal medulla can be considered as a modified version of a postganglionic sympathetic nerve fibre.

Once released, the hormones of the adrenal medulla are inactivated by the same processes as those responsible for inactivation of the noradrenaline released by the neurones of the sympathetic nervous system; thus there may be some reuptake into the postganglionic sympathetic neurones where the transmitters are stored in granules for reuse, or the transmitters may be degraded enzymatically. The principal enzymes involved in the inactivation of adrenaline and noradrenaline are monoamine oxidase (MAO) and catechol-O-methyltransferase (COMT), the latter being the more important. The actions of these enzymes result in the formation of the metabolites metadrenaline, normetadrenaline, 3-methoxy-4-hydroxymandelic acid (vanillyl mandelic acid, VMA) or a small amount

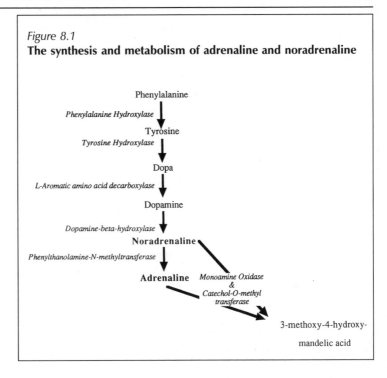

Figure 8.1
**The synthesis and metabolism of adrenaline and noradrenaline**

of 3-methoxy-4-methoxyphenylglycol (MHPG) which are secreted in the urine (see Figure 8.1). Small amounts of adrenaline and noradrenaline are excreted unchanged in the urine.

## 8.3 Actions of adrenaline and noradrenaline

Adrenaline and noradrenaline both act on the same receptors, known collectively as adrenoceptors. These receptors are membrane-bound, with seven-transmembrane domains, and are linked to G-proteins (see Chapter 1). Adrenoceptors are divided into several different subtypes, each of which mediates specific actions. The earliest classification divided the adrenoceptors into $\alpha$-adrenoceptors and $\beta$-adrenoceptors; $\alpha$-adrenoceptors have approximately the same sensitivity to adrenaline and noradrenaline while $\beta$-adrenoceptors are much more sensitive to adrenaline than to noradrenaline. These subtypes are then further divided into $\alpha_1$-adrenoceptors and $\alpha_2$-adrenoceptors and $\beta_1$-, $\beta_2$- and $\beta_3$-adrenoceptors. The $\alpha_1$-adrenoceptors are found in the smooth muscle of arterial blood vessels, the genitourinary system, the eye, and the smooth muscle associated with hair follicles. They are also associated with salivary and sweat glands and with those liver cells responsible for glycogenolysis. The action of the catecholamines on this receptor subtype is mediated by the $G_q$ form of the G-protein which results in the stimulation of membrane bound phospholipase C and the production of the second messengers inositol triphosphate ($IP_3$) and diacylglycerol (DAG). Stimulation of the $\alpha_1$-adrenoceptor generally results in smooth muscle contraction and an increase in glycogen-olysis and glandular secretion.

The $\alpha_2$-adrenoceptors of the central nervous system are often presynaptic and act as part of a feedback process to inhibit the further release of the neurotransmitter noradrenaline. There are, however, some specific actions associated with $\alpha_2$-adrenoceptors. In the periphery they are present on the gastrointestinal tract and on the $\beta$ cells of the pancreas where they act to produce relaxation of the gut and to decrease insulin secretion. When $\alpha_2$-adrenoceptors are activated the $G_2$ protein inhibits the activity of adenyl cyclase and therefore reduces the synthesis of the second messenger cAMP. In veins the effect of the $G_2$ protein is to open a calcium ion channel which results in an influx of calcium, causing venoconstriction.

Stimulation of both $\beta_1$- and $\beta_2$-adrenoceptors results in the activation of the $G_s$ form of the G-protein. The result of this is a stimulation of adenyl cyclase and an increase in the production of cAMP. $\beta_1$-adrenoceptors predominate in the heart where their stimulation causes an increase in the rate and force of contraction. The $\beta_1$-adrenoceptors are also responsible for the lipolytic effects of adrenaline, although there may be involvement of $\beta_3$-adrenoceptors. The predominant form of $\beta$-adrenoceptor in the blood vessels of skeletal muscle and the smooth muscle of the bronchial tree is the $\beta_2$-subtype whose stimulation results in vasodilatation and bronchodilatation respectively.

The primary effect of the secretion of the adrenal medullary hormones is to prepare the body for a stressful event. Thus the initial response to any stressful event is activation of the sympathetic nervous system, the effects of which are mediated predominantly by noradrenaline (see above). This nervous response, typically, is relatively short lived; however one feature of the response is to cause secretion of adrenaline and noradrenaline by the adrenal medulla. This has the effect of not only prolonging the responses initiated by the sympathetic nervous system activation, but also switching the response from that predominated by noradrenaline (i.e. $\alpha$-adrenoceptor activity) to one predominated by adrenaline (i.e. increasing the $\beta$-adrenoceptor component). This response has often been labelled the 'fight, flight or fright' response (see Box 8.1).

Probably the most important effects of the catecholamines are those on the cardiovascular system. Stimulation of the $\alpha_1$-adrenoceptors results in constriction of the blood vessels supplying the skin and gastrointestinal tract while stimulation of $\beta_2$-adrenoceptors results in dilatation of the blood vessels supplying the skeletal muscle; hence there is a shift of blood flow away from the gut and skin towards the skeletal muscle. The effect of $\beta_1$-adrenoceptor stimulation is to increase the rate and force of contraction of the heart which results in the increased distribution of oxygen and nutrients. Because of the combined effects of the vasoconstriction and the vasodilatation, and because of the presence of other autoregulatory mechanisms, the catecholamines normally have little effect on systemic blood pressure.

The receptors for adrenaline and noradrenaline are subdivided into $\alpha_1$-, $\alpha_2$-, $\beta_1$-, $\beta_2$- and $\beta_3$-adrenoceptors.

Adrenaline is secreted in response to stress and produces the 'fight or flight' response.

*Box 8.1*  **The actions of adrenaline and noradrenaline**

Bronchodilatation
Increased lipolysis
Increased mental alertness
Relaxation of the gastrointestinal tract
Increased blood flow to brain and muscle
Increased rate and force of contraction of the heart
Decreased blood flow to skin and gastrointestinal tract

The increased distribution of oxygen is aided by the dilatation of the bronchi and bronchioles, which is mediated by $\beta_2$-adrenoceptors. The increased distribution of nutrients is a result of the increased availability of carbohydrate as a result of the increased lipolysis which follows stimulation of the $\beta$-adrenoceptors and the increased glycogenolysis induced by $\alpha_1$-adrenoceptor stimulation. Paradoxically $\alpha_1$-adrenoceptor stimulation also results in the inhibition of insulin secretion, although this action is over-ridden by the effects of the raised blood glucose.

All of these effects prepare the body for activity in that they provide all of the requirements for increased skeletal muscle activity while reducing the requirements of other tissues such as the gut and skin. Other effects of the catecholamines which facilitate the response to stressful stimuli are the relaxation of the gut, thereby decreasing extraneous energy utilization, dilatation of the pupil of the eye, thereby increasing visual acuity, and increased central nervous stimulation.

Prolonged, serious stress results in the stimulation of the adrenal cortex after which the corticosteroids surplant the catecholamines as the hormonal response to stress (see Chapter 9).

## 8.4 Disorders of the adrenal medulla

The adrenal medulla can be removed without there being any apparent deterioration of the general health of the individual and no clinical symptoms occur following damage of the gland as a result of tuberculosis or a tumour. This gland is therefore not essential for life.

The adrenal medulla can be removed without any deleterious effect on health.

The most important disorder of the adrenal medulla is phaeochromocytoma, which is a tumour which causes increased, but often sporadic, secretion of adrenaline and noradrenaline. It is a very rare condition which occurs in less than 1 in 100 000 of the population, predominantly between the ages of 20 and 60; in 10 per cent of cases the tumours are bilateral. Typical symptoms would be short-lived (10–15 minute) attacks of panic, sweating, pallor, chest pain and severe headache which may be precipitated by factors such as emotion, exercise or some foodstuffs. There may be postural hypotension although there are also cases of hypertension. It is estimated that phaeochromocytoma may account for as many as 1 in 500–1000 cases of hypertension. A potentially fatal hypertensive

Phaeochromocytoma
is characterized by
recurrent panic attacks.

crisis may be precipitated by anaesthesia or surgery. The usual form of treatment is surgical removal of the tumour. This treatment is generally effective although it should be noted that the surgery required for the treatment of the condition may act to precipitate a fatal episode of that condition. This problem may be overcome by use of pharmacological antagonists for $\alpha$- and $\beta$-adrenoceptors to control the blood pressure either prior to or during the anaesthesia and surgery (see Box 8.2).

Box 8.2   **The treatment of phaeochromocytoma**

---

Phaeochromocytoma is caused by excessive secretion of adrenaline which then acts on adrenoceptors to produce the characteristic symptoms of sweating, pallor, headache, postural hypotension or hypertension. Many of these symptoms can be controlled by the combined use of antagonists of $\alpha$-adrenoceptors and $\beta$-adrenoceptors.

| Drug | Proprietary name |
|---|---|
| *α-adrenoceptor antagonist* | |
| Phenoxybenzamine | Dibenyline (capsules) |
| | |
| *β-adrenoceptor antagonists* | |
| Propranolol | Inderal (tablets) |
| Acebutolol | Sectral (capsules) |
| Atenolol | Tenormin (tablets) |
| Labetalol | Trandate (tablets) |
| Metoprolol | Betaloc (tablets) |
| Pindolol | Visken (tablets) |

Phaeochromocytoma may also be treated with metirosene (proprietary name: Demser) which inhibits tyrosine hydroxylase, and therefore prevents the synthesis of adrenaline and noradrenaline.

---

# Summary

- The adrenal glands lie above the kidneys and are divided into the adrenal cortex and the adrenal medulla which secretes adrenaline and noradrenaline. The adrenal medulla can be considered as a continuation of the sympathetic nervous system, and is therefore stimulated by any form of sympathetic nervous system activation, such as stress.

- The effects of the secreted adrenaline and noradrenaline have been labelled the 'fight, flight or fright response'. Thus following secretion of these hormones there is an increase in the rate and force of contraction of the heart, broncho-dilatation, liberation of glucose from fats, and enhanced blood flow to the muscles, at the expense of blood flow to the skin and gut. Adrenaline and noradrenaline also increase mental awareness.

- There are no clinical conditions associated with underactivity of the adrenal medulla, but overactivity, possibly due to secretory tumours, is associated with recurrent panic attacks and may be important in many cases of hypertension.

# Selected reading

Hadley, M.E., 1992, Catecholamines and the sympathoadrenal system. In Hadley, M.E., *Endocrinology*, 3rd edn, Englewood Cliffs: Prentice-Hall, 362–390

Laycock, J. and Wise, P., 1996, The adrenal medulla. In Laycock, J. and Wise, P., *Essential Endocrinology*, 3rd edn, Oxford: Oxford University Press, 134–144

# 9 The Adrenal Cortex

## 9.1 Introduction

As described in the previous chapter, the adrenal glands lie above the kidneys and are divided into two distinct portions, the inner adrenal medulla (10 per cent) and the outer adrenal cortex (90 per cent). The adrenal cortex synthesizes many different hormones which are involved in the control of sodium/potassium balance and carbohydrate metabolism and which play a vital role in the body's reponse to stresses such as injury, infection and emotion. All of the hormones of the adrenal cortex share a similar chemical structure, being derived from cholesterol; these hormones are collectively known as steroid hormones. Unlike the adrenal medulla, a functioning adrenal cortex is essential for life.

---

A functioning adrenal cortex is essential for life.

---

## 9.2 Synthesis of adrenocortical steroid hormones

The adrenal cortex is divided into three distinct layers, the outer zona glomerulosa, the middle zona fasciculata and the innermost zona reticularis. Cells within the different areas possess different ranges of enzymes and are therefore concerned with the synthesis of different adrenocortical hormones (see later). All of the hormones synthesized by the adrenal cortex are derived from cholesterol which is obtained predominantly from the diet but may be synthesized within the gland itself, and is stored as lipid droplets within the cytoplasm of the adrenocortical cells. The various hormones of the adrenal cortex are not stored to any great extent, thus their secretion is dependent upon the rate of their synthesis. Because of their common origin, all of the adrenocortical hormones share a common chemical nucleus, the cyclopentanoperhydrophenanthrene, or steroid, ring; the structure of a typical steroid, oestradiol, is presented in Chapter 1, Figure 1.4. The synthetic pathway for these hormones is shown in Figure 9.1; a similar pathway exists for the synthesis of the sex hormones within the testes and ovaries although the relative activities of the enzymes involved in the synthesis differs, thus altering the relative proportions of the final products (see Chapters 10 and 11). Steroid hormones produce a variety of effects, but they are usually classified according to their predominant action, thus the major secretions of the adrenal cortex are the glucocorticoids and the mineralocorticoids which are

Figure 9.1
**The synthetic pathway for the production of adrenal steroid hormones from cholesterol**

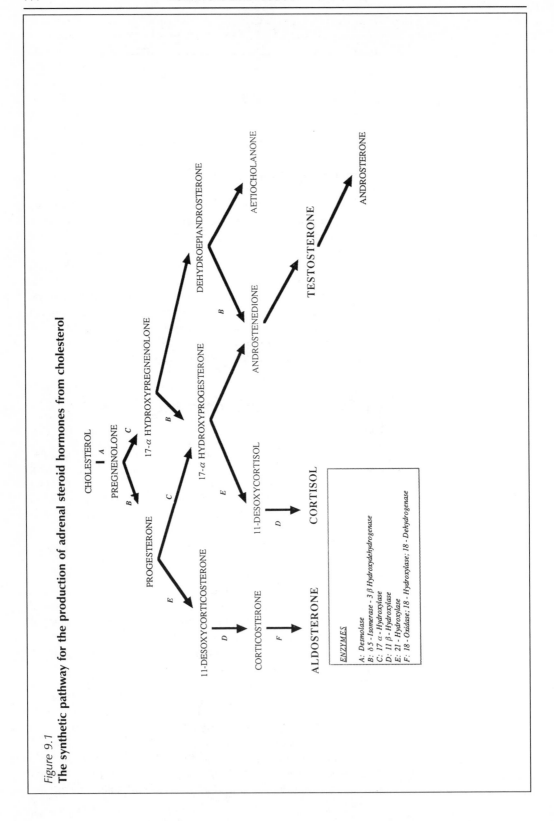

concerned with the control of carbohydrate metabolism and sodium/potassium balance respectively. By far the most important representative of the mineralocorticoids is aldosterone, while cortisol and corticosterone are the most important hormones with glucocorticoid activity. The 18-hydroxylase enzyme involved in the synthesis of aldosterone is found only within the zona glomerulosa, thus this hormone is synthesized only in that region of the adrenal cortex, while the 17α-hydroxylase enzyme required for the production of 17α-hydroxypregnenolone, 17α-hydroxyprogesterone and the hormones derived from them is found only in the zona fasciculata and the zona reticularis; thus cortisol is synthesized in the zona fasciculata and androgens (see below) are synthesized in the zona reticularis. Under normal circumstances the adrenal cortex also secretes small quantities of male sex hormones such as dehydroepiandrosterone, androstenedione and testosterone, which are collectively called the androgens, and the female sex hormone oestradiol, which is a member of the oestrogens. The quantities of the sex hormones secreted are relatively small compared with the amounts secreted by the gonads, thus they are of little physiological significance, however disorders of the adrenal cortex may increase their secretion thus resulting in developmental or reproductive abnormalities.

> Under normal circumstances the adrenal gland secretes insignificant amounts of androgens; androgens become important at times of adrenal disorder.

In chemical terms the different groups of adrenocortical steroids can be differentiated by the number of carbon atoms contained, thus glucocorticoids and mineralocorticoids both contain 21 carbon atoms, androgens contain 19 and oestrogens contain 18 carbon atoms. Physiologically, however, the delineation is not so complete, for example many of the oestrogens and androgens exert some glucocorticoid and mineralocorticoid activity. The synthesis, secretion and actions of the mineralocorticoids and the glucocorticoids are controlled independently, therefore these two groups of hormones will be considered separately; the adrenocortical androgens and oestrogens are only of significance in disorders of the adrenal cortex and are therefore discussed later.

## 9.3 Secretion and actions of glucocorticoids

### 9.3.1 *Control of glucocorticoid secretion*

The most important naturally occurring glucocorticoids are cortisol and corticosterone; cortisol will be used as the example of a glucocorticoid. The conversion of cholesterol to pregnenolone and the resultant synthesis and secretion of cortisol is stimulated by adrenocorticotrophic hormone (ACTH) from the anterior pituitary gland. The secretion of ACTH itself is stimulated by corticotrophin releasing factor (CRF) from the hypothalamus (see Chapter 2). The secretion of ACTH is pulsatile, with a marked diurnal pattern: the peak of secretion is in the early morning, at the time of waking, and the nadir is in the middle of the night. This pattern is related to sleep–wake patterns, and is therefore disrupted by shift work and long-haul travel. Cortisol secretion shows the same pattern but the peak and nadir occur approximately two hours later than those of ACTH. Secretion of ACTH is also raised at times of prolonged stress.

ACTH produces its effect by interaction with a membrane bound receptor on cells of the adrenal cortex; this interaction results in the synthesis of the second messenger cAMP. It is cAMP that stimulates the various enzymes involved in the synthesis of cortisol but the enzymes for synthesis of adrenal androgens and oestrogens are also stimulated; thus ACTH has an effect on the secretion of these other hormones. The secretion of ACTH is under negative feedback control via an effect of cortisol on the anterior pituitary gland. Because ACTH also acts as a growth factor for the adrenal cortex, absence of ACTH results in adrenal cortical atrophy while excess ACTH causes adrenal cortical hyperplasia.

### 9.3.2  *Actions of glucocorticoids*

Only 10 per cent of cortisol within the blood is in a free, active form; the remainder is bound to the plasma proteins corticosteroid binding globulin (CBG or transcortin) (75 per cent) and albumin (15 per cent). The same proteins also transport the other glucocorticoids and progesterone. Pregnancy is associated with an increase in CBG which results in a compensatory increase in circulating plasma cortisol concentration although the amount of free cortisol remains stable. Metabolism of the adrenal steroids occurs mainly in the liver where they are glucuronidated to form water soluble forms which are excreted in the urine.

In common with all other steroid hormones, glucocorticoids produce their effects by an action on intracellular receptors and alterations in gene expression. The unbound receptor resides in the cytoplasm, but it migrates to the nucleus following interaction with the glucocorticoid. Cortisol thus influences protein synthesis and in so doing influences the activity of a range of intracellular enzymes (see Chaper 1). This mechanism of action inevitably results in a delay, in the order of hours or days, between the secretion of the hormone and the manifestation of its effects. In some cases however, the effects of cortisol are rapid, for example the feedback inhibition of ACTH secretion, it has therefore been postulated that cortisol may produce some of its effects by directly altering the properties of the target cell membrane.

At normal physiological concentrations the most important actions of cortisol are those on carbohydrate metabolism. Cortisol acts generally in a manner opposite to that of insulin (Chapter 7), thus cortisol antagonizes the effects of insulin on cellular uptake of glucose, it stimulates glycogenolysis and it stimulates hepatic gluconeogenesis. Cortisol also stimulates lipolysis and mobilization of fatty acids, partially by potentiating the effects of growth hormone and the catecholamines, although in excessive concentrations cortisol causes fat synthesis and deposition in novel anatomical sites, most notably the face, the trunk and the intrascapular region of the

**Cortisol acts in a manner opposite to insulin to elevate blood glucose concentrations.**

shoulders (see later). In the liver, cortisol stimulates amino acid uptake as a precursor to enhanced gluconeogenesis, but in the periphery it inhibits amino acid uptake and protein synthesis which, when coupled with the enhanced gluconeogenesis, can result in a net loss of skeletal protein.

Glucocorticoids are also able to stimulate aldosterone receptors and thus influence excretion of sodium and potassium. However aldosterone-sensitive tissues possess an enzyme, $11\alpha$-hydroxy-steroid dehydrogenase, which converts cortisol to inactive cortisone; thus at normal concentrations the mineralocorticoid actions of glucocorticoids are limited, although at high concentrations the actions may become apparent.

Another effect of excess cortisol is an enhanced vasoconstrictor response to catecholamines which results in increased blood pressure. This in turn increases glomerular filtration rate in the kidney which is accompanied by an increased excretion of calcium. The increased excretion of calcium is also related to the inhibitory effects of the glucocorticoids on 1,25-dihydroxycholecalciferol (see Chapter 5). This results in decreased calcium reabsorption by the kidney and decreased calcium uptake in the gut. The resultant decrease in plasma calcium induces parathyroid hormone secretion and a promotion of bone readsorption in an attempt to liberate calcium. This effect, coupled with an inhibition of calcitonin activity and a direct effect on bone formation means that glucocorticoids reduce bone density. Glucocorticoids can also produce psychological effects with possible feelings of elation or sedation.

At times of psychological and physiological stress such as trauma, infection or hypoglycaemia, there is a rapid secretion of ACTH and a resultant increase in the concentrations of circulating corticosteroids. At these raised concentrations additional effects of these hormones become apparent. One of the most important of these effects is an inhibitory effect on the body's defence systems. Within the body cellular damage results in the liberation of inflammatory mediators such as 5-hydroxytryptamine and bradykinin and the synthesis and release of other inflammatory mediators such as the prostaglandins and leucotrienes which are produced by the action of the enzyme phospholipase $A_2$ on the phospholipid component of the damaged cell's membrane. These inflammatory mediators increase the permeability of the neighbouring capillary wall, which leads to diffusion of fluid into the damaged area, and exert a chemotactic effect which attracts white blood cells (leucocytes). In turn, the leucocytes release proteolytic enzymes which remove the cell debris before collagen formation completes the process of tissue repair.

A similar inflammatory response may occur following exposure of the body to a foreign protein or an infecting organism. The presence of the foreign protein stimulates a response by the lymphoid tissue of the immune system. This response may be cell mediated, or mediated by the production of antibodies (see Figure 9.2). The antibodies produced by the B lymphocytes may inactivate viruses or bacterial toxins, or may induce the death of the invading cells by activation of the complement system or by making the cells easier

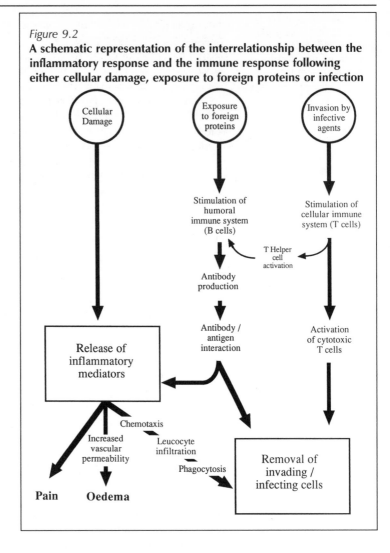

Figure 9.2

**A schematic representation of the interrelationship between the inflammatory response and the immune response following either cellular damage, exposure to foreign proteins or infection**

targets for phagocytosis. An antigen–antibody reaction may also cause the rupture of mast cells/basophils resulting in the release of histamine. In a manner similar to the inflammatory mediators mentioned previously, histamine not only increases vascular permeability but also induces symptoms typical of allergic reactions such as itch. The cell mediated immune process involves T lymphocytes which are able to kill the invading cells (cytotoxic T cells) and which are able to enhance antibody production by the B cells (Helper T cells).

Because the activity of the immune system often results in damage to the invading cells, it is often accompanied by an activation of the inflammatory response.

Glucocorticoids affect these defence systems at many levels:

- They suppress the lymphoid tissue, and therefore reduce the antibody production and inhibit the cellular immune system.

- They stabilize leucocyte membranes and therefore reduce the release of proteolytic enzymes.

- They inhibit phospholipase $A_2$ and therefore reduce the synthesis of the inflammatory mediators.

The glucocorticoids thus reduce the immune response by several mechanisms. Paradoxically, however, certain chemical mediators of the immune response, for example interleukin 1, induce corticosteroid secretion, by enhancing the release of CRF and ACTH. It has therefore been suggested that the corticosteroids may be involved in a negative feedback process to limit the extent of the immune response.

In the absence of corticosteroids even mild stress can be fatal.

As described above, there are many effects of gluocorticoids, but their most important role is in the response to stress. In the absence of corticosteroids even mild stress can be fatal. In this response to stress, however, many of the effects of these hormones, at first glance, appear paradoxical. In any individual the initial response to stress is activation of the sympathetic nervous system followed rapidly by secretion of adrenaline by the adrenal medulla (Chapter 8). This results in an increase in the rate and force of contraction of the heart, dilatation of the bronchioles and redistribution of blood flow from the skin and gut to the brain and muscles. There is also an enhanced mobilization of glucose. All of these are features of the fright, fight or flight response in which all of the requirements necessary for a sudden increase in physical activity are made available. The response of the adrenal medulla to the stressful stimulus is, however, of limited duration; thus at times of prolonged stress the adrenal cortex takes over control of the stress response. The glucocorticoids maintain the enhanced supply of glucose which may be required for the prolonged response to the stressor but also initiate additional effects such as suppression of the inflammatory response.

The purpose of the inflammatory and immune responses is to limit the extent of any tissue damage. The pain alerts the sufferer to the damage while the oedema, a consequence of the increased permeability of the capilliaries, has the effect of diluting any toxic substances that may be present. In the case of joints, the oedema also causes immobilization which aids stabilization. The infiltration by leucocytes allows the destruction of any invading cells while antibodies inactivate the foreign proteins. Tissue repair is enhanced by prostaglandins. In preventing the processes of the inflammatory and immune responses it would appear that the glucocorticoids potentiate the adverse effects of the injury and retard tissue repair. The steroids thus appear to be acting contrary to the individual's best interests.

The value of the adrenocortical stress response is that it decreases the inflammatory response, thus it removes the pain and decreases the immobilization induced by the oedema; the steroid-induced sedation also causes a lack of awareness of the severity of the situation. The overall effect is that the individual is able to perform despite the presence of the injury or

The glucocorticoids inhibit the inflammatory response and therefore allow the individual to perform in spite of his or her injuries.

infection. In nature this effect is of paramount importance, and it is because of these properties that glucocorticoids are used in the relief of many illnesses.

## 9.4 Secretion and actions of mineralocorticoids

### 9.4.1  *Control of mineralocorticoid secretion*

Aldosterone and 11-deoxycorticosterone are the physiologically important mineralocorticoids; aldosterone will be used as the example. Unlike the secretion of cortisol, the secretion of aldosterone is relatively uninfluenced by ACTH, although ACTH stimulates the initial conversion of cholesterol to pregnenolone. The major controlling factor in the secretion of aldosterone is the renin–angiotensin system (see Figure 9.3). Renin is an enzyme produced by the juxtaglomerular cells of the kidney in response to the decreased renal blood supply that may accompany a decrease in systemic blood pressure, fluid loss or haemorrhage; renin catalyses the conversion of angiotensinogen, which is formed in the liver, to angiotensin I. Angiotensin I acts as a precursor for the formation of angiotensin II which, together with its metabolite angiotensin III,

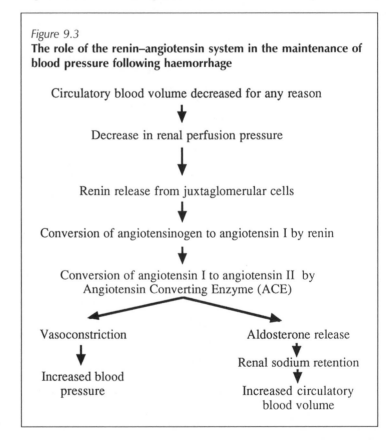

*Figure 9.3*
**The role of the renin–angiotensin system in the maintenance of blood pressure following haemorrhage**

Circulatory blood volume decreased for any reason

Decrease in renal perfusion pressure

Renin release from juxtaglomerular cells

Conversion of angiotensinogen to angiotensin I by renin

Conversion of angiotensin I to angiotensin II  by Angiotensin Converting Enzyme (ACE)

Vasoconstriction

Increased blood pressure

Aldosterone release

Renal sodium retention

Increased circulatory blood volume

acts on G-protein linked receptors of the zona glomerulosa to induce the synthesis and secretion of aldosterone.

Aldosterone is also influenced by atrial natriuretic peptide (ANP), a peptide produced by the heart. When the atrial muscle is stretched it secretes ANP which inhibits aldosterone secretion and the formation of renin. The result of the actions of ANP is thus to reduce sodium reabsorption and to decrease blood volume. The secretion of aldosterone is also directly stimulated by trauma, anxiety, hyperkalaemia and hyponatraemia.

### 9.4.2  *Actions of mineralocorticoids*

Within the circulation aldosterone is only 50 per cent protein bound and like cortisol, has specific intracellular receptors. The interaction of aldosterone with its receptors in the target tissues results in the production of novel proteins which enhance the transport of sodium and potassium ions across the cell membrane. Aldosterone thus stimulates the reabsorption of sodium ions in the distal tubule of the kidney, with some lesser effects in the collecting duct, proximal tubule and ascending loop of Henlé, and in the colon, sweat and salivary glands. The sodium reabsorption occurs in exchange for either potassium or hydrogen ions, thus aldosterone causes excretion of potassium ions. By controlling reabsorption of sodium ions, aldosterone influences plasma sodium concentration, which in turn influences water reabsorption in the collecting duct via an effect on ADH secretion (Chapter 3). The interaction of the renin–angiotensin system, aldosterone and ADH therefore controls blood volume and influences blood pressure.

> Aldosterone influences blood volume by controlling reabsorption of sodium ions in the kidneys.

## 9.5 Disorders of adrenocortical activity

### 9.5.1  *Hypoactivity*

Decreased activity of the adrenal cortex may be due to a disease of the gland itself, or a disorder of the secretion of ACTH or CRF by the anterior pituitary or hypothalamus respectively. Diseases of the gland would normally result in decreased secretion of both glucocorticoids and mineralocorticoids, while disorders of the hypothalamic–pituitary axis only influence the glucocorticoids; it is very rare for there to be decreased aldosterone secretion without an accompanying decrease in glucocorticoids.

In many cases the symptoms of adrenocortical hypoactivity go unnoticed for many years, whereas in other cases there is very rapid onset that may be fatal. The lack of aldosterone results in decreased blood pressure resulting from excessive sodium and fluid loss, although other compensation mechanisms mean that the blood pressure may remain within normal limits. The lack of glucocorticoids causes hypoglycaemia and thus symptoms of tiredness and weakness. In the rare cases where the onset of the disease is rapid the presenting symptoms may be hypotension, nausea, vomiting,

> Addisonian crisis may result in death within 24 hours.

diarrhoea, confusion and coma which may culminate in death within 24 hours. If the disease is slower in onset there may be a single symptom of tiredness or a more extensive condition with tiredness, weakness, anorexia and weight loss. Prolonged deficiency of glucocorticoids may be accompanied by an increase in the secretion of ACTH due to the release of the anterior pituitary gland from negative feedback. Increased ACTH produces skin pigmentation reminiscent of a sun tan, however unlike a true sun tan, the pigmentation is greatest at pressure points such as belts or straps and there may be marked pigmentation of the soles of the feet and the buccal cavity. A deficiency of glucocorticoid secretion is called Addison's disease which has a prevalence of approximately 3 per 100 000 of the population. This condition, which is often caused by an autoimmune destruction of the adrenal cortex and in one third of cases follows tuberculosis, usually presents in adults and is more common in females than in males. A rapid onset of glucocorticoid insufficiency is called Addisonian crisis. Addison's disease may occur alongside Graves' disease (Chapter 4), hypogonadism (Chapters 10 and 11) and in diabetes mellitus (Chapter 7) as part of an autoimmune multiple endocrine deficiency.

Treatment of the glucocorticoid deficiency is by administration of a glucocorticoid such as hydrocortisone or, more rarely, by administration of an ACTH analogue. In the first instance, treatment of mineralocorticoid deficiency is by infusion of saline to replace the fluid and sodium, possibly followed by treatment with a mineralocorticoid such as fludrocortisone (see Box 9.1). With the correct treatment, the prognosis of Addison's disease is very good: the skin pigmentation fades, a normal lifespan is achieved and normal pregnancies may be undertaken by females.

### 9.5.2  *Hyperactivity*

Increased secretion of aldosterone may occur as a consequence of a tumour within the zona glomerulosa (primary hyperaldosteronism or Conn's syndrome), which occurs in less than 0.2 per cent of the population and accounts for 1 per cent of cases of hypertension, or is a consequence of increased activity of the renin–angiotensin system, for example due to renal artery stenosis, which may be involved in up to 10 per cent of all cases of hypertension. Apart from the effects on the systemic blood pressure as a consequence of the increase in angiotensin II, the symptoms of the two conditions are similar. Conn's syndrome usually develops between the ages of 30 and 60, more commonly in females, although occasionally bilateral adrenal hyperplasia presents in 10- to 20-year-olds. The predominant symptom of hyperaldosteronism is hypertension resulting from sodium and fluid retention coupled with hypokalaemia which induces muscle weakness. The usual form of treatment for Conn's syndrome is the surgical removal of the tumour, although in some cases the condition can be controlled by use of the aldosterone receptor antagonist spironolactone. In secondary hyperaldosteronism the condition can be relieved by use of spironolactone or an angiotensin II antagonist such as losartan.

Increased secretion of glucocorticoids is referred to as Cushing's syndrome. This is a relatively common endocrine disorder which usually presents between the ages of 30 and 40, and more commonly in females. The condition first described by Cushing was associated with raised ACTH secretion, therefore this form of the disorder is often referred to as Cushing's disease. Of the cases of Cushing's syndrome, 75 per cent are due to inappropriate ACTH secretion, of which one third are due to a secretory tumour of the pituitary gland and about 10 per cent are due to ectopic secretion of ACTH by small cell tumours of the lung, C-cell tumours of the thyroid gland or phaeochromocytoma. In 20 per cent of cases the Cushing's syndrome is due to a tumour of the adrenal cortex. It has recently been postulated that some forms of the disease may be due to the existence of a constituitively active CRF receptor on the anterior pituitary gland.

The symptoms of Cushing's syndrome reflect the multiple actions of the glucocorticoids. Thus there is hyperglycaemia resulting in polyuria and polydipsia and a loss of muscle mass due to protein usage for gluconeogenesis. The limbs therefore appear thin, but the deposition of new fat stores around the abdomen, face and shoulders causes enlargement of the torso. Protein removal from bone has the consequence of osteoporosis, while removal of protein from the structure of the skin results in the appearance of characteristic stretch marks, or striae, across the abdomen. If the condition occurs in children the abnormal protein metabolism results in disorders of growth. Other actions of the glucocorticoids also become apparent at these high circulating concentrations, thus the mineralocorticoid effects induce sodium and fluid retention resulting in hypertension, and androgenic effects cause acne and excessive hair growth with menstrual disturbances in women. In Cushing's disease the raised ACTH results in orange skin pigmentation. The characteristic features of the condition have variously been described as 'moon face and buffalo hump' because of the skin pigmentation together with the abnormal fat depostion around the face and shoulders, or 'oranges on sticks' because of the above mentioned features coupled with the wasting of the limbs and the fat deposition around the torso. If left untreated, Cushing's syndrome usually proves fatal within five years, death being due to cardiovascular disease, complications of diabetes mellitus or infection.

The usual treatment of Cushing's syndrome is the surgical removal of the tumour, although destruction of the affected gland is sometimes achieved using external irradiation. Drugs are available to inhibit the normal synthesis of the adrenal steroids by inhibition of either 11$\beta$-hydroxylase or 3$\beta$-hydroxysteroid dehydrogenase, but these usually only provide temporary relief because of an increased secretion of ACTH (see Box 9.1).

**Box 9.1    Drugs used in the treatment of disorders of the adrenal cortex**

*Hypoactivity*
Underactivity of the adrenal cortex results in a deficiency of glucocorticoids and/or
mineralocorticoids; the treatment of these conditions is thus by replacement therapy with orally
active synthetic corticosteroids.

|                     | *Drug*          | *Proprietary name*     |
|---------------------|-----------------|------------------------|
| *Mineralocorticoid* | Fludrocortisone | Florinef (tablets)     |
| *Glucocorticoid*    | Hydrocortisone  | Hydrocortone (tablets) |

*Hyperactivity*
Excess secretion of glucocorticoids (Cushing's syndrome) or mineralocorticoids (Conn's syndrome)
may be treated by surgical removal of the source of the hormones although pharmacological
therapies are also available. These therapies act either by blocking the action of the hormone at
its receptor or by preventing the conversion of cholesterol to aldosterone and cortisol.

|                                                       | *Drug*                | *Proprietary name*      |
|-------------------------------------------------------|-----------------------|-------------------------|
| *Aldosterone receptor antagonists*                    |                       |                         |
|                                                       | Spironolactone        | Aldactone (tablets)     |
|                                                       | Spironolactone        | Spriroctan (tablets)    |
|                                                       | Potassium canrenoate  | Spiroctan-M (injection) |
| *11β-hydroxylase inhibitor*                           |                       |                         |
|                                                       | Metyraprone           | Metopirone (capsules)   |
| *δ5-isomerase-3-β-hydroxydehydrogenase inhibitor*     |                       |                         |
|                                                       | Trilostane            | Modrenal (capsules)     |

### 9.5.3    *Congenital adrenal hyperplasia*

A group of conditions exist which are caused by an inborn error of
adrenocorticosteroid synthesis; these are known collectively as congenital adrenal hyperplasia. They are rare, with a total incidence of
approximately one per 5000 births, but a discussion of their symptoms and treatment serves as an excellent illustration of the range of
symptoms that may occur in disorders of the adrenal cortex (see
Table 9.1).

In congenital adrenal hyperplasia there is an inborn error in one
or more of the enzymes required for the synthesis of cortisol.
Because of the deficiency in cortisol secretion the anterior pituitary
gland is not subject to feedback control and therefore secretes
abnormally high amounts of ACTH. As a result of the raised ACTH
there is increased uptake of cholesterol by the adrenal cortex and
enlargement of the adrenal cortex (hyperplasia).

In the case of a deficiency of the desmolase enzyme, there is
no metabolism beyond cholesterol thus the gland is unable to
synthesize any of the adrenocorticosteroids and because the same
synthetic pathway exists they are unable to synthesize the sex steroids. This condition results in feminization of external genitalia of
male infants because of the deficiency of androgens (see Chapter 10)
and usually results in death. Treatment would require extensive

**Table 9.1** The effects of enzyme deficiency on the secretion of adrenocorticosteroids and the features of congenital adrenal hyperplasia

| Enzyme deficiency | Hormone deficiency | Clinical features |
|---|---|---|
| Desmolase | All | Hypoadrenalism – usually results in death Feminization of external genitalia in male infant |
| 3-$\beta$-hydroxydehydro-genase | Cortisol Aldosterone | Salt loss Ambiguous external genitalia in males Virilization in females |
| 17-$\alpha$-hydroxylase | Cortisol Androgens | Hypertension Hypokalaemia No secondary sexual development in females Pseudohermaphroditism Gynaecomastia in males |
| 21-hydroxylase | Cortisol Aldosterone | Virilization of females Possible salt loss |
| 11-$\beta$-hydroxylase | Cortisol Aldosterone | Hypertension Virilization of females |
| 18-hydroxylase | Aldosterone | Salt loss |
| 18-dehydrogenase | Aldosterone | Salt loss |

hormone replacement therapy with glucocorticoids, mineralocorticoids and the appropriate sex hormones.

A deficiency of 3$\beta$-hydroxydehydrogenase means that synthesis of cortisol and aldosterone is impossible. Because of the lack of cortisol there is increased secretion of ACTH and therefore hypertrophy of the adrenal cortex and increased uptake of cholesterol. The only route of metabolism still possible for the cholesterol is to dehydroepiandrosterone, a weak androgen. The symptoms of this condition are therefore those of Addison's disease due to the deficiency of cortisol, salt loss due to deficiency of aldosterone and virilization of the external genitalia of female infants because of the accumulation of the weak androgen, but ambiguous external genitalia in males because of the inability to synthesize the potent androgen testosterone (see Chapters 10 and 11).

A deficiency of 17-$\alpha$-hydroxylase occurs in about 1 in 14 000 births and is characterized by deficiencies of cortisol and the androgens; the excess cholesterol is thus metabolized to the mineralocorticoids. The resultant symptoms are those of Conn's syndrome with hypertension and hypokalaemia, together with the symptoms of Addison's disease. The inability to synthesize the sex hormones results in pseudohermaphroditism in male infants with lack of development of secondary sexual characteristics in either sex (see Chapters 10 and 11).

A deficiency of either 21-hydroxylase, which is the most common of the disorders of this type, occurring in 1 in 10 000 births, or 11-$\beta$-hydroxylase, causes a deficiency of cortisol, resulting in Addison's disease, and a deficiency of aldosterone resulting in the possibility of salt loss. The excess cholesterol is metabolized to produce androgens such as testosterone. In the male this condition may result in precocious puberty but it may cause virilization of females. Deficiencies of 18-hydroxylase or 18-dehydrogenase result in a decreased synthesis of aldosterone and a resultant salt loss.

In many cases the abnormality of the enzyme concerned is not absolute, thus there is a decrease in the production of the appropriate hormones rather than a total lack of synthesis. Because of this, the conditions may remain undetected for many years with the only symptoms being hirsutism or menstrual disturbances; frequently the condition is only recognized during investigations of infertility. The treatment for congenital adrenal hyperplasia invariably involves administration of the deficient hormones, which also has the effect of providing negative feedback thereby reducing the ACTH secretion and resolving the excess production of the other hormones. In milder cases, the management of the condition has been so successful that some females treated from early childhood have successfully borne children.

## 9.6 Pharmacological uses of adrenocorticosteroids

Glucocorticoids are used in replacement therapy for the treatment of conditions such as Addison's disease or congenital adrenal hyperplasia, for their immunosuppressive or anti-inflammatory effects in the treatment of conditions such as arthritis, asthma or allergies, or for the treament of proliferative conditions such as leukaemia. Mineralocorticoids are used only for replacement therapy.

The shorth plasma half-life of aldosterone renders it unsuitable for mineralocorticoid replacement therapy, thus the drug of choice is fludrocortisol. In most cases the selection of the glucocorticoid is similarly dependent upon the pharmacokinetics of the available agents and the predominant effect required. Most glucocorticoids are orally active but their absorption through the skin varies as does their plasma half-life. For replacement therapy the natural hormones cortisol and cortisone, which must undergo conversion to cortisol, are the drugs of choice because they have limited immunosuppressive activity but high mineralocorticoid activity. In the treatment of inflammatory conditions prednisolone, methylprednisolone and prednisone, which is converted to prednisolone, are of use because they have good anti-inflammatory/immunosuppressant effects with minimal mineralocorticoid actions. Dexamethasone and betamethasone are also very potent immunosuppressants with minimal mineralocorticoid activity, but their long plasma half-life and their growth-suppressing properties render them unsuitable for long-term use; they are therefore only used for the treatment of severe, acute inflammatory disorders (see Box 9.2).

Many adverse effects are associated with the use of glucocorticoids, all related directly to the known actions of cortisol. Steroid usage may suppress wound healing and may exacerbate infections

Box 9.2 **Currently available drugs which are based on adrenal steroids**

Corticosteroids are used widely to control the symptoms of a variety of disorders which involve an inflammatory or an immune response. These drugs may be used either systemically (in the form of tablets or injections) or locally (in the form of creams, inhalers or nasal sprays). All of the corticosteroids used are potent glucocorticoids with little or no mineralocorticoid or androgenic actions.

| Drug | Relative potency* | Proprietary name |
|---|---|---|
| Hydrocortisone | 1 | Efcortelan (cream); Hydrocortisyl (cream); Canesten HC (cream); Daktacort (cream) |
| Prednisolone | 5 | Precortisyl Forte (tablet); Prednesol (tablet) |
| Methylprednisolone | 6 | Medrone (tablets, injection) |
| Triamcinolone | 6 | Kenalog (injection); Ledercort (tablets) |
| Betamethasone | 30 | Betnelan (tablets); Betnovate (cream) Betnesol (tablets, injection, drops) |
| Dexamethasone | 30 | Decadron (tablets, injection) |
| Beclomethasone | 125 | AeroBec (inhaler); Becotide (inhaler); Beconase (nasal spray); Propaderm (cream) |
| Budesonide | 125 | Pulmicort (inhaler); Rhinocort (nasal spray); Preferid (cream) |
| Fluticasone | 250 | Flixotide (inhaler); Flixonase (nasal spray); |
| Flunisolide | 250 | Syntaris (nasal spray) |

*Approximate potency; higher values represents greater potency. No account is taken of the duration of action.

due to their immunosuppresant effects. Long-term use in children may cause inhibition of growth, and in adults may result in osteoporosis. The development of diabetes mellitus and other symptoms of Cushing's syndrome also often accompanies steroid therapy. Probably the most important adverse effect, however, is suppression of the hypothalamic–pituitary axis. Chronic administration of exogenous glucocorticoids results in suppression of ACTH secretion which in turn leads to atrophy of the adrenal cortex. If steroid therapy is then stopped abruptly, the adrenal cortex is unable to secrete endogenous hormones and the patient suffers an Addisonian crisis, which may be fatal. This consequence can be overcome by the gradual reduction of the dose of the exogenous steroid in order to allow regeneration of the adrenal gland. Other consequences of suppression of the anterior pituitary may include disturbances of sex hormone secretion resulting in symptoms such as menstrual disturbances.

It should always be remembered that the use of corticosteroids for suppression of the inflammatory response has the effect of

Chronic administration of adrenocortical steroids leads to atrophy of the adrenal cortex.

'Steroids allow the patient to walk to the mortuary.'

removing the symptoms without affecting the cause of the underlying disorder, which may worsen. This property of these steroids is the reason for the sentiment that 'steroids enable the patient to walk to the mortuary'.

## Summary

- The adrenal gland secretes glucocorticoids, mineralocorticoids and androgens, all of which share a common steroid structure. The most important of these are the glucocorticoids and the mineralocorticoids, which are essential for the maintenance of life; the androgens are normally produced in physiologically insignificant quantities, but their actions may become apparent in adrenal cortex disorders.

- The secretion of the mineralocorticoids is controlled by plasma sodium/potassium balance and by the renin–angiotensin system, which itself is influenced by blood pressure. Once secreted the mineralocorticoids act on the kidney to promote sodium reabsorption, in exchange for potassium or hydrogen ions. A deficiency of mineralocorticoids results in salt loss, while an excess causes sodium retention, fluid retention and hypertension.

- The secretion of the glucocorticoids is controlled by adrenocorticotrophic hormone (ACTH) from the anterior pituitary, and corticotrophin releasing factor (CRF) from the hypothalamus, both of which are influenced by negative feedback. The major physiological actions of glucocorticoids are to raise plasma glucose concentrations by enhancing glycogenolysis, lipolysis and gluconeogenesis. At times of stress the secretion of the glucocorticoids is increased which causes other actions of these hormones to become apparent, most notably an inhibition of the normal inflammatory and immune responses. A deficiency of glucocorticoids is characterized by an inability to cope with stress, accompanied by hypoglycaemia. Excess glucocorticoids induce hyperglycaemia, achieved partially by utilization of sketelal proteins, and an increased susceptibility to infection.

- The glucocorticoids are used therapeutically to inhibit the inflammatory and immune responses, and therefore they reduce the symptoms of a variety of illnesses.

## Selected reading

Alberts, B., Bray, D., Lewis, J., Raff, M., Roberts, K. and Watson, J.D., 1994, The cellular basis of immunity. In Alberts, B., Bray, D., Lewis, J., Raff, M., Roberts, K. and Watson, J.D., *The Molecular Biology of the Cell*, 3rd edn, New York: Garland Publishing, 1196–1206

Binkley, S.A., 1995, Glucocorticoids and the adrenal cortex. In Binkley, S.A., *Endocrinology*, New York: HarperCollins Publishers, 282–300

Binkley, S.A., 1995, Aldosterone and adrenals, angiotensin and kidneys. In Binkley, S.A., *Endocrinology*, New York: HarperCollins Publishers, 301–317

Laycock, J. and Wise, P., 1996, The adrenal cortex. In Laycock, J. and Wise, P., *Essential Endocrinology*, 3rd edn, Oxford: Oxford University Press, 99–133

Margioris, A.N., Gravanis, A. and Chrousos, G.P., 1994, Glucocorticoids and mineralocorticoids. In Brody, T.M., Larner, J. and Neu, H.C. (eds), *Human Pharmacology: Molecular to Clinical*, 2nd edn, St Louis: Mosby, 473–481

New, M.I. and White, P.C., 1995, Genetic disorders of steroids synthesis and metabolism. In Thakker, R.V. (ed), *Baillière's Clinical Endocrinology and Metabolism, International Practice and Research. Vol. 9, No. 3, Genetic and Molecular Biological Aspects of Endocrine Disease,* London: Baillière Tindall, 525–554.

Schimmer, B.P. and Parker, K.L., 1996, Adenocorticotrophic hormone; adrenocortical steroids and their synthetic analogues; inhibitors of the synthesis and actions and adrenocortical hormones. In Hardman, J.G., Linbird, L.E. and Gilman, A.G. (eds), *Goodman and Gilman's The Pharmacological Basis of Therapeutics*, 9th edn, New York: McGraw-Hill, 1459–1485

# 10 The Testes and the Male Reproductive System

## 10.1 Introduction

The most important hormone of the male reproductive system is testosterone, the definitive example of the group of hormones called the androgens. As described in Chapter 9, androgens are synthesized and secreted by the adrenal cortex in both males and females; in males however the major source of androgens are the Leydig cells of the testes. Plasma testosterone concentrations in women are approximately 0.5–2.5 nmol/l (0.2–0.7 ng/ml), all of which is secreted from the adrenal cortex and the ovaries; concentrations in men are approximately 10–30 nmol/l (3–9 ng/ml), of which 95 per cent is secreted from the testes. In adult males the synthesis and secretion of testosterone by the Leydig cells is controlled by luteinizing hormone (LH) secreted from the anterior pituitary gland, the secretion of which, in turn, is controlled by gonadotrophin releasing hormone (GnRH) from the hypothalamus. Testosterone exerts a negative feedback effect to inhibit the secretion of both LH and GnRH (see Figure 10.1).

## 10.2 Synthesis and mechanism of action of testosterone

The pathway for the synthesis of testosterone by the Leydig cells is similar to that of the adrenal cortex, thus it involves the conversion of cholesterol firstly to pregnenolone and thence to dehydroepiandrosterone (DHEA), although a small portion of the pregnenolone is converted to androstenedione via progesterone. Both DHEA and androstenedione are ultimately converted to testosterone. Unlike the adrenal cortex, the Leydig cells lack the enzymes necessary for the conversion of cholesterol to the mineralocorticoids or the glucocorticoids. The cholesterol is normally obtained from the plasma low density lipoprotein components, although it can be synthesized by the Leydig cells *de novo*; the conversion of cholesterol to pregnenolone is the rate limiting step and is under the control of LH. Once secreted, approximately 98 per cent of the plasma testosterone is protein-bound: 40 per cent to sex hormone binding globulin (SHBG), 40 per cent to albumin and the remainder to other proteins. Inactivation of testosterone occurs in the liver and results in the excretion of the glucuronide and sulphate conjugates of androsterone and aetiocholanone in the urine.

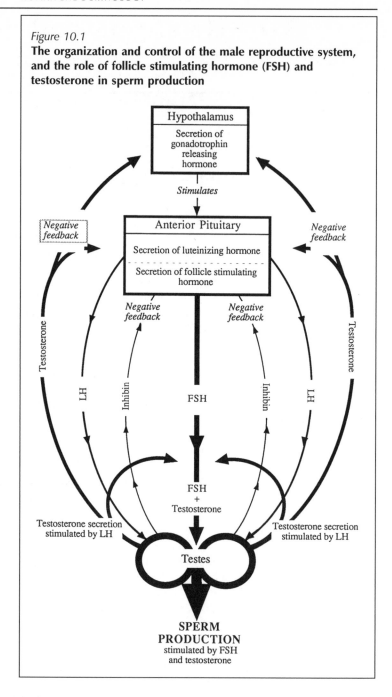

Figure 10.1

**The organization and control of the male reproductive system, and the role of follicle stimulating hormone (FSH) and testosterone in sperm production**

Like the other steroid hormones, testosterone produces its biological effects by interaction with specific receptors within either the nucleus or the cytoplasm of the target cell. Stimulation of the receptor causes changes in transcriptional or translational processes within that cell resulting in alterations of protein synthesis. Many androgen-sensitive tissues are able to enhance the activity of testosterone by first converting it to the more potent androgen dihydro-testosterone (DHT), thus increasing the biological effects. The enzyme responsible for this conversion is $5\alpha$-reductase which is found in the cytoplasm. In a small number of tissues the testosterone must first be converted to $17\beta$-oestradiol by aromatase before it can produce its effects via oestrogen receptors (see later). Androgens such as testosterone are responsible for the development and maintenance of the characteristic male physique and for male fertility, thus they have a wide range of long-lasting effects and they are important throughout life, but some of their most important effects begin as early as seven weeks after conception.

**Androgens control the development of male physique and fertility.**

## 10.3 The effects of androgens *in utero*

At the point of fertilization the genetic sex of the embryo/foetus is determined by the chromosomes of the gametes; the maternal gametes can only contribute X sex chromosomes while the paternal gametes can contribute either X or Y sex chromosomes. Fusion of an ovum with a spermatozoon possessing an X chromosome will result in the development of a female embryo while fusion of an ovum with a spermatozoon possessing a Y chromosome will result in the development of a male embryo. It is the Y chromosome which determines the sex of the embryo as absence of a Y chromosome always results in the development of a female while even in the presence of multiple X chromosomes, the presence of a Y chromosome will result in the development of a male.

**The presence of a Y chromosome will always result in the development of male primary sexual characteristics.**

At the early stages of development the embryo possesses gonads that have the potential to develop into either ovaries or testes. These gonads are composed of separate cortical and medullary regions but there is an endogenous tendency for the development of the medulla to be suppressed and that of the cortex to be enhanced. Such development eventually gives rise to an ovary. In the presence of a Y chromosome, or a fragment of that chromosome called the Sry gene, however, the cortical development is suppressed while that of the medulla is enhanced which gives rise to a testis. Testicular development normally begins at about the seventh week of gestation, while ovarian development does not begin until about week 13. Ovarian development, therefore, is often described as being the 'default option', which means that it will occur unless there is prior activation of the Sry gene, although complete differentiation of the embryonic ovaries requires the presence of two X chromosomes.

**Development of female anatomy is the 'default option'.**

Associated with the undifferentiated embryonic gonads are two structures, called the Müllerian duct and the Wolffian duct, which have the potential to develop into the Fallopian tubes, uterus and vagina or the epididymis, vas deferens and seminal vesicles respectively. At about the seventh week of gestation, the embryonic gonads begin to convert cholesterol to pregnenolone and thence to the active sex hormones. In a female embryo the pregnenolone is converted, via testosterone, to $17\beta$-oestradiol, but in a male embryo the presence of the Sry gene of the Y chromosome suppresses the production of aromatase and thus prevents the conversion of cholesterol beyond testosterone. It is the testosterone that is secreted by the embryonic male gonads that, among other things, causes further differentiation of the Wolffian duct into the epididymis, vas deferens and seminal vesicles (see Figure 10.2). Note that at this stage the testicular synthesis and secretion of testosterone does not require the secretion of LH by the embryonic anterior pituitary because of the availability of cholesterol derivatives of placental origin. In addition to stimulating the secretion of testosterone, the presence of the Sry gene also causes the production of a glycoprotein called Müllerian inhibiting factor (MIF, also called Müllerian regression factor or antiMüllerian hormone) by the Sertoli cells of the embryonic testes. MIF exerts a local effect to induce atrophy of the Müllerian duct and therefore to prevent the development of female internal genitalia.

The testosterone secreted from the embryonic testes also causes the development of the masculine infantile external genitalia, an action which requires the conversion of testosterone to DHT.

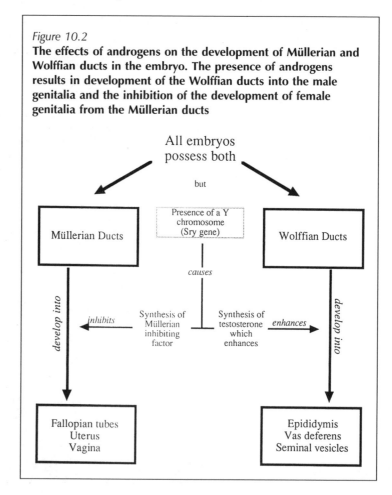

*Figure 10.2*

**The effects of androgens on the development of Müllerian and Wolffian ducts in the embryo. The presence of androgens results in development of the Wolffian ducts into the male genitalia and the inhibition of the development of female genitalia from the Müllerian ducts**

Development of the penis and scrotum begins shortly after the onset of Wolffian duct differentiation and the testes begin to descend into the scrotum from about the seventh or eighth month of gestation (see Figure 10.3). This descent, which is dependent upon the presence of both testosterone and MIF, is normally complete by the time of birth but up to 10 per cent of males have undescended testes at birth, with up to 1 per cent having undescended testes at the age of five years.

At birth, 10 per cent of males have undescended testes.

There is also evidence that the testosterone secreted by the foetal or neonatal testes produces changes in brain function that persist throughout life. In male rats, neonatal castration causes a change in the hypothalamus such that the later secretion of the gonadotrophins by the anterior pituitary after puberty (see later) is cyclical in a pattern similar to that seen in normal female rats (see Chapter 11). Conversely treatment of female rats with testosterone shortly after birth abolishes the normal cyclical pattern of gonadotrophin and sex hormone secretion. There is a critical period, which in rats is at about the time of birth, when this effect of testosterone occurs, but once the changes have taken place they are irreversible. In a similar manner, there is a prenatal critical period in rats in which exposure to testosterone causes the rats to exhibit male pattern mating behaviour (i.e. mounting) in later life irrespective of genetic sex. Again, deficiency of testosterone at this period can lead to expression of female behaviour (i.e. lordosis) irrespective of sex. This latter effect of testosterone is of particular interest because it has been shown that the response is dependent upon stimulation of oestrogen receptors. It has been demonstrated that the brain contains the aromatase enzyme and is therefore able to convert testosterone to 17$\beta$-oestradiol; it is this oestradiol that induces the development of the male sexual behaviour that is manifested in later life. The effect is not induced by DHT and can be blocked by

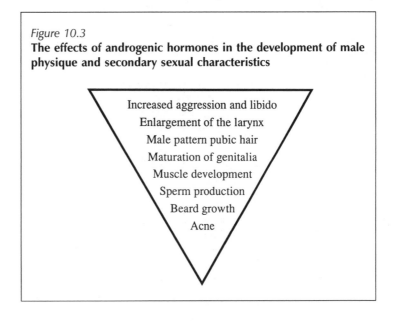

Figure 10.3
**The effects of androgenic hormones in the development of male physique and secondary sexual characteristics**

Increased aggression and libido
Enlargement of the larynx
Male pattern pubic hair
Maturation of genitalia
Muscle development
Sperm production
Beard growth
Acne

oestrogen receptor antagonists and by inhibitors of 5$\alpha$-reductase. Paradoxically the effect does not occur in female rats because the ovarian secretion of oestrogens is insufficient to allow adequate 17$\beta$-oestradiol to reach the appropriate brain areas. Whether such behavioural effects are important in humans remains unclear.

## 10.4 Disorders of development of male primary sexual characteristics

A lack of androgens results in the genetically male baby being labelled and reared as a girl.

All of the aforementioned demonstrates that an embryo or foetus will develop female internal and external genitalia and female brain function unless it is exposed to the actions of testosterone or some other androgen. It therefore follows that in the absence of such effects a genetic male embryo would develop, and be born with, female genitalia, and that because gender is normally assigned to the baby at birth dependent upon the external genitalia, the baby would be brought up as a female, irrespective of the true genetic make-up.

A disorder or imbalance of testicular function in the foetus may result in hermaphroditism. A true hermaphrodite is an individual who possesses both ovarian and testicular tissue. Such a condition is extremely rare in humans, but in laboratory animals it has been shown that unilateral castration of a male embryo results in the development of a female genital tract (Fallopian tube and uterus) on the side from which the testis was removed and a male genital tract on the contralateral side; there would be ambiguous external genitalia (see later) because of the reduced secretion of testosterone. In persistent Müllerian duct syndrome there is a failure of the testes to produce MIF although testosterone secretion is normal. In these cases there is normal development of the male external and internal genitalia but these males also possess Fallopian tubes and a uterus. This is an example of male pseudohermaphroditism: male because of the presence of the testes but pseudo- because there is no female gonadal tissue. Another example of a male pseudohermaphrodite would be one in whom the embryonic testes are unable to synthesize or secrete testosterone, for example in cases of a genetic lack of one of the enzymes required for its synthesis as in some forms of congenital adrenal hyperplasia (see Chapter 9). In these cases the MIF causes atrophy of the Müllerian ducts but the lack of androgens causes a failure to develop normal male external or internal genitalia. Although the external genitalia are unambiguously female and the clitoris normal, the vagina is short and blind-ended and the internal genitalia are absent except for testes which may be found either within the abdomen or within the labia majora. Similar features are seen in individuals in whom there is a genetic lack or abnormality of androgen receptors when the condition is referred to as testicular feminization.

It is important to realize that the enzyme or receptor abnormality may be such that there is a reduced secretion of testosterone, or reduced response, rather than a complete failure. In these individuals there may be development of ambiguous external genitalia. Typically in such cases the external genitalia are predominantly female, however there may be a testosterone induced enlargement of the clitoris, giving it a penis-like appearance, with partial fusion of the labia. As described above, the testes may be found within the folds of the labia. In most cases the treatment of such conditions involves the removal of the source of the androgens and a surgical remodelling of the genitalia to that of a female.

In a small number of individuls there is an inborn deficiency of $5\alpha$-reductase. In these individuals those tissues in which testosterone is normally converted to DHT fail to develop, while those that are normally responsive to testosterone develop normally. In these individuals therefore there is atrophy of the Müllerian duct and development of the Wolffian duct, but the external genitalia develop as a vagina with scrotum-like labia in which may be found the testes, similar to that described above. The specific features of this condition relate to the fact that those tissues in which testosterone is converted to $17\beta$-oestradiol or in which testosterone produces its effects without prior conversion to DHT, develop normally. These specific features become more prominant at puberty (see later).

## 10.5 Puberty and spermatogenesis

In human males, the secretion of testosterone falls drastically after birth and the Leydig cells become quiescent until the time of puberty. The age of puberty differs between different ethnic groups and can be affected by factors such as nutrition, but usually occurs at about 12–14 years. The stimulus for the onset of puberty is unknown but it may be related to the sensitivity of the prepubertal hypothalamus to circulating androgens. During the prepubertal years the adrenal cortex is the only source of androgens, but it is believed that the sensitivity of the hypothalamus is such that these low levels of circulating androgens are sufficient to inhibit GnRH secretion by negative feedback. Puberty commences with the secretion of LH and FSH by the anterior pituitary gland. This is believed to be a consequence of a sudden decrease in the sensitivity of the hypothalamus to negative feedback by the adrenal androgens. The factors involved in the timing of this event are unknown but some workers have suggested that body weight may be important, this proposal being based on the clinical observation that puberty may be delayed in cases of malnutrition or anorexia nervosa and early in obese individuals.

Puberty commences with an increase in the secretion of LH and FSH from the anterior pituitary gland.

The effect of the increased secretion of FSH and LH is to increase the number and activity of Sertoli and Leydig cells respectively within the testes; thus the first sign of puberty in males is testicular enlargement. The adult testes are each about 4 cm long and 2 cm wide and their primary functions are the production of spermatozoa and the secretion of testosterone. Each testis consists of densely coiled seminiferous tubules in which the spermatozoa are produced. Lining the seminiferous tubules are the Sertoli cells. These cells perform several functions, such as the secretion of nutrients and other factors that are required by the developing spermatozoa; they also perform a phagocytotic role and they form a 'blood–testes' barrier which prevents the passage of large proteins (for example antibodies) from the blood to the lumen of the tubule. The seminiferous tubules are connected, via the rete testis to the epididymis and the vas deferens. Lying between the seminiferous tubules are the interstitial or Leydig cells which are responsible for the synthesis and secretion of testosterone and small amounts of $17\beta$-oestradiol and progesterone, although the role of the latter two hormones in the development and control of the male reproductive system is unknown.

Initiation of spermatogenesis requires both FSH and testosterone. It has been suggested that the testosterone from the Leydig cells enters the seminiferous tubules and is taken up by the Sertoli cells which have produced an androgen-binding protein in response to stimulation by FSH. The presence of the testosterone within the Sertoli cell enables the full maturation of the spermatocytes to spermatids and finally spermatozoa. FSH cannot initiate or maintain spermatogenesis alone, but in some cases it has been demonstrated that spermatogenesis can be maintained by testosterone alone, in the absence of FSH, provided that it was originally initiated at puberty by FSH. The secretion of FSH by the anterior pituitary is controlled independently of LH secretion by a feedback mechanism involving the production of a peptide, called inhibin, by the Sertoli cells. Inhibin acts on either the hypothalamus or the anterior pituitary to limit FSH secretion.

> Initiation of spermatogenesis requires both FSH and testosterone.

## 10.6 Effects of testosterone and the development of secondary sexual characteristics

As well as being characterized by the initiation of spermatogenesis, male puberty is also a time of increased secretion of testosterone. As described above, the first sign of puberty is the testicular enlargement induced by FSH and LH. It is due to LH that testosterone secretion by the testes is stimulated; testosterone is responsible for all of the other changes that occur at puberty.

Those sex hormone dependent changes that occur at the time of puberty are referred to a secondary sexual characteristics; these characteristics are superimposed upon the primary sexual characteristics that developed *in utero*. One of the first testosterone dependent secondary sexual characteristics that develops at puberty is the maturation of the external genitalia. Testosterone causes enlargement of the scrotum to accommodate the developing testes and growth of the penis. There is also an increase in the content of the endoplasmic reticulum within the cells of the prostate gland, seminal vesicles and the bulbourethral glands. This results in the secretion of a dilute fluid rich in calcium citrate, fibrinolysin, acid phosphatase and prostaglandins by the prostate gland, a seminal fluid rich in fructose by the seminal vesicles and mucus by the bulbourethral glands. The combination of these secretions forms the thick seminal fluid which is responsible for the nutrition, transport and maintenance of viable, motile spermatozoa.

The androgens also cause growth of male pattern pubic hair. The typical male pattern of hair growth in the pubic area is that of a triangle of hair with the apex uppermost and a male escutcheon growing towards the navel. There is also facial hair growth and growth in the axillae; at later stages of puberty there may also be growth of chest hair. Paradoxically the androgens are also responsible for the temporal recession and baldness that may occur in later life, although it is probable that this effect only occurs in those men with a genetic disposition to baldness. Another effect of the androgens is to increase the activity of the sebaceous glands; this action is responsible for the acne commonly seen in males at about the time of puberty.

Puberty is also characterized by a spurt in linear growth. This is caused by the androgens, in conjunction with growth hormone, causing an increase in the rate of protein synthesis within the long bones of the arms and the legs. The growth spurt in males typically commences at the age of 13 and results in an increase in height of about 20 cm by the age of 17. This action results in the shift from the typical juvenile proportions with relatively short limbs to the more adult proportions with an increase in limb length to trunk length ratio. The androgens are also responsible for initiating the closure of the epiphyses which leads to the cessation of growth at about the age of 17.

The effects of the androgens on protein synthesis are also responsible for the pubertal enlargement of the larynx which causes the deepening of the voice, and the growth of skeletal muscle. In skeletal muscle the androgens act to stimulate an increase in protein synthesis and therefore an increase in muscle mass; this action is referred to as an anabolic effect, thus androgens are anabolic steroids. Androgens also stimulate erythropoiesis both by direct stimulation of bone marrow and by increasing the synthesis of erythropoietin by the kidney. These effects explain the greater body weight and strength of males compared to females (see Figure 10.3).

Androgens also have psychological effects leading to increased libido and increased aggression.

> Androgens cause growth of facial hair, but are also responsible for baldness.

## 10.7 Disorders of testosterone secretion

Excessive testosterone in an adult male normally goes unrecognized because of the wide variation in the extent of androgen activity seen in the normal population. Such conditions can occur, however, in cases of testicular or adrenal tumours or in individual misuse of therapeutic androgens for their anabolic actions (see later). The usual presenting symptom would be that of decreased fertility, possibly with some testicular atrophy. These symptoms arise because of the suppression of gonadotrophin secretion by the anterior pituitary.

Excessive or inappropriate secretion of androgens in young males may produce precocious puberty. Onset of puberty before the age of nine is considered abnormal. If the increased testosterone secretion is a result of testicular stimulation by gonadotrophins released as a consequence of a hypothalamic or pituitary tumour the condition is referred to as true precocious puberty. In this condition there will be enlargement of the testes, sperm production and development of secondary sexual characteristics. If the precocious puberty is a consequence of exogenous androgens, excessive androgen secretion by the adrenal cortex – for example congenital adrenal hyperplasia – or an androgen secreting tumour, there will be development of secondary sexual characteristics without testicular development or sperm production. Such a condition is referred to as pseudoprecocious puberty. Precocious puberty may be arrested by administration of an androgen receptor antagonist such as cyproterone acetate and by removal of the source of the hormone secretion; however the effects cannot be reversed, thus it is usually necessary to counsel the child and to reassure them that their classmates will go through the same changes within a few years. One of the features of excess androgen activity before or at the time of puberty is an initially rapid spurt in linear bone growth followed by early closure of the epiphyses. The individual is therefore taller than his classmates at the start of puberty, but by the age of 17 is shorter and stockier than his contemporaries.

Androgen secreting tumours may also occur in females. The effects of the androgens in females would be to cause enlargement of the clitoris with development of male pattern pubic hair (including facial hair) and an increase in muscle mass. There would also be a deepening of the voice and excessive activity of the sebaceous glands; there may also be increases in libido and aggression. The actions of the androgens on the anterior pituitary gland would cause disruption of the menstrual cycle. If the condition remains untreated it may result in baldness.

The effects of reduced androgen secretion, or reduced responses to androgens *in utero* have been described previously. The effects of a reduction of androgen secretion in childhood or adulthood depend upon the age of onset. Such conditions may arise because of testicular failure, castration or because of a disorder of either the hypothalamus or the anterior pituitary gland. The presence of high concentrations of oestrogens, either from an oestrogen secreting tumour or from an exogenous source also decreases testosterone production due to inhibition of gonadotrophin secretion.

If the deficiency occurs before puberty there will be a lack of, or a delay in, the development of the male secondary sexual characteristics. There is no spurt in linear bone growth, but similarly there is no closure of the epiphyses, thus the individual is initially shorter than his classmates, but eventually attains a height greater because the growth continues beyond the age of 18. The features of this condition are described as eunuchoid because there is no increase in muscle mass, enlargement of the larynx or growth of pubic hair. The external genitalia remain infantile and there is no sperm production. One cause of male hypogonadism and delayed or failure of puberty is Klinefelter's syndrome which occurs in 0.2 per cent of all males. This condition is characterized by an extra X chromosome, thus the genotype is XXY, although in some cases there may be XXXY or XXYY, which is termed mosaicism. In some cases the only presenting symptom is infertility, while in others there is failure to develop at puberty. Gynaecomastia (development of breasts in males, see Chapter 2) is also a common feature.

If the deficiency starts after puberty there is a decrease or cessation of sperm production with loss of muscle mass and pubic hair. There is also a decrease in libido and aggression. There is little or no change in voice. Treatment of these conditions may necessitate androgen replacement therapy, or if fertility and sperm production is required, administration of gonadotrophins.

In those genetic males in whom an inability to secrete or respond to testosterone *in utero* lead to the development of female external genitalia (see above), the same disorder usually causes the failure of testosterone secretion at puberty. There is therefore a failure to develop male secondary sexual characteristics. There will however be no development of female secondary sexual characteristics (see Chapter 11), thus juvenile female anatomy is retained. This condition may be treated by administration of female sex hormones to initiate development of female secondary sexual characteristics (see Chapter 11). One exception to this is $5\alpha$-reductase deficiency. As described previously this condition results in the development of male internal genitalia, with female external genitalia. At puberty those features dependent upon testosterone activity develop, thus there is increased muscle mass, deepening of voice and male pattern pubic hair growth, but those features dependent upon DHT remain unaffected, although there may be some enlargement of the clitoris. There is less facial hair and acne and the prostate gland fails to develop. These individuals normally retain their feminine gender assignment.

# 10.8 Pharmacological uses of androgens

When used therapeutically testosterone is usually administered by injection within an oil vehicle. This route of administration avoids the first pass metabolism by the liver and provides sustained androgen activity. Orally active synthetic androgenic/anabolic steroids such as stanozolol have also been produced. As described above, androgens may be used for hormone replacement therapy in the treatment of hypogonadism in males. They are also sometimes used for their anabolic actions in the treatment of certain wasting disorders.

Anabolic androgens have also been misused by athletes in an attempt to increase muscle mass and strength. There is some evi-

Anabolic androgens
may be misused by
athletes to enhance
performance.

dence of their efficacy, however it is possible that their actions owe
more to their glucocorticoid activity than to the anabolic/andro-
genic actions. As described in Chapter 9, steroid hormones have
the ability to suppress inflammation and therefore to limit the
symptoms of tissue damage. It has been suggested that the use of
steroids by athletes enables them to continue training for longer,
being less affected by fatigue or injury. This action, together with
an increased aggression induced by the androgens, may be responsible for the improved performance seen among users of
these steroids.

The side-effects of anabolic steroid abuse are predictable. In
adult males the only manifest side-effect is infertility and testicular
atrophy due to the suppression of gonadotrophins. This effect may
be irreversible. There may also be increased risks of cardiovascular
disease and prostate carcinoma. In adult females the use of anabolic
steroids results in disruption of the menstrual cycle, clitoral enlargement and development of male pattern pubic hair. If the
drugs are used in children they can cause precocious development
of male secondary sexual characteristics in both males and females.

Antagonists of the androgen receptor may also be used therapeutically, for example in the treatment of disorders related to excess
secretion or response to these hormones (see Box 10.1).

## 10.9 Male hormonal contraceptives

The aim of a male contraceptive is to suppress sperm production in
a reversible manner. As already described, initiation of sperm production is dependent upon the presence of both FSH and testosterone, but once initiated sperm production can, in some individuals,
be maintained by testosterone alone. Several regimes have been
tested as male contraceptives (see Figure 10.4). The most obvious
is the administration of an androgen receptor antagonist such as
cyproterone acetate. This has the effect of reducing sperm count,
but also causing regression of some of the male secondary sexual
characteristics, including a decrease in libido. These side-effects
render this form of contraception unusable. Later attempts investigated the use of GnRH analogues which suppress the secretion of
LH and FSH due to down regulation of the GnRH receptors. There
was a marked decrease in sperm production, but the inhibition of
LH secretion induced testosterone deficiency causing regression of
secondary sexual characteristics. The suppression of FSH secretion
also causes testicular atrophy. Some of these problems can be overcome by administration of the drug danazol which is an inhibitor of
LH and FSH secretion but also possesses slight androgenic properties. When given in combination with testosterone this treatment
caused azoospermia or severe oligospermia in 85 per cent of
subjects; no severe side-effects were reported. Further attempts to
suppress FSH secretion have used medroxyprogesterone in combination with testosterone replacement. Of the subjects tested, 60 per
cent became azoospermic with a further 30 per cent becoming

*Box 10.1* **The use of hormones and their analogues and antagonists in the treatment of prostatic carcinoma**

The prostate gland is an example of a tissue that develops and grows under the influence of hormones; other exampes of such tissues are the testes, the ovaries and the breasts. In these tissues an abnormal response to the hormones may result in uncontrolled cellular replication: cancer. Prostate cancer is thus an example of a hormone dependent cancer and most forms of treatment aim to deprive the tumour of those hormones required for its continued growth, in this case androgens. The actions of the hormones on the tumour can be decreased in several ways:

- Removal of the source of endogenous hormones.
- Prevention of the synthesis or secretion of the hormones.
- Use of a pharmacological antagonist of the hormones.
- Use of a physiological antagonist of the hormones.

In the case of prostate cancer, the source of the androgens is the testes, although the adrenal glands also secrete small amounts of androgens. A standard treatment for prostate cancer is therefore castration, although this does not influence secretion of androgens by the adrenal cortex. The secretion of androgens by the testes can also be reduced by the administration of GnRH analogues. These drugs mimic the actions of endogenous GnRH and therefore initially stimulate the secretion of LH and FSH and ultimately testosterone. After about ten days of use, however, the GnRH receptors down regulate which results in complete inhibition of LH and FSH secretion.

The effects of androgens can also be reduced by the administration of androgen receptor antagonists. These drugs have the advantage that they also reduce the effects of androgens of adrenal origin, but the disadvantage that they remove the effects of negative feedback, and therefore may induce an increased secretion of androgens by the testes. An alternative approach is to use high dose oestrogen therapy; this form of treatment has the effect of reducing the influence of the androgens, by acting in an opposing manner, but carries with it a wide range of adverse effects dependent upon the oestrogenic effects, for example gynaecomastia (see Chapter 2) and increases risk of thromboembolism (see Chapter 11).

*Available drugs*

|                      | Drug          | Proprietary name                    |
|----------------------|---------------|-------------------------------------|
| GnRH analogues       | Buserelin     | Suprefact (injection, nasal spray)  |
|                      | Goserelin     | Zoladex (implant)                   |
|                      | Leuprorelin   | Prostap (injection)                 |
|                      | Triptorelin   | De-capeptyl (injection)             |
| Androgen antagonists | Cyproterone   | Cyprostat (tablets)                 |
|                      | Flutamide     | Drogenil (tablets)                  |
|                      | Bicalutamide  | Casodex (tablets)                   |
| Oestrogens           | Fosfestrol    | Honvan (tablets, injection)         |
|                      | Polyestradiol | Estradurin (injection)              |

oligospermic. The trials were discontinued because of the fears of progestogenic and oestrogenic side-effects.

The most extensive trial of a male hormonal contraceptive to date involved 271 men being given testosterone oenanthate by weekly intramuscular injection. Of these men 157 (58 per cent) became azoospermic. The partners of these 157 men were requested to discontinue their contraception and one pregnancy occurred; this is equivalent to 0.8 pregnancies per 100 person years among responders, a rate comparable with that of the female oral contraceptive.

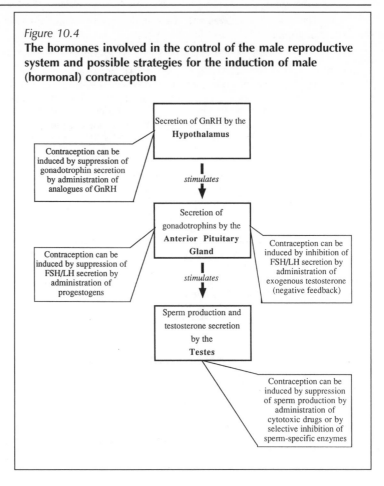

Figure 10.4

**The hormones involved in the control of the male reproductive system and possible strategies for the induction of male (hormonal) contraception**

---

Male hormonal contraceptives may be as effective as female oral contraceptives.

The mean time to azoospermia was 120 days with a mean time to recovery of normal sperm count of 90 days; 3.7 per cent of the volunteers had an increase in body weight, 5.7 per cent had raised haemoglobin and 16.5 per cent had decreased testicular volume. There was a drop out rate of 3 per cent due to acne, 1 per cent due to increased aggression and libido, and 0.7 per cent due to abnormal lipid profile.

To date this is the most effective form of reversible male hormonal contraceptive, but it must be remembered that it produced azoospermia in only 58 per cent of the subjects, and it was administered by weekly intramuscular injection, a method which is likely to induce poor user satisfaction.

Other attempts at male contraception have investigated oral agents which inhibit sperm activity. Probably the best known of these is gossypol which is derived from cotton seed. Following clinical trials in which there was an initial daily loading dose with monthly maintenance it has been claimed that gossypol is 99 per cent effective in causing azoospermia or oligospermia. There were, however, reports of side-effects such as fatigue (12.8 per cent), gastrointestinal disturbance (8.4 per cent), decreased libido

(6 per cent) and severe hypokalaemia (less than 5 per cent). Recovery of sperm production was slow, with irreversible azoospermia in 10 per cent of subjects. The mechanism of action of gossypol is unknown although inhibition of sperm specific enzymes, testicular degeneration and chelation of important ions have all been suggested. Trials of gossypol have ceased due to fears concerning toxicity, carcinogenicity, mutagenicity, and teratogenicity.

Another attempt at male contraception has used 6-chloro-6-deoxysugars which are taken up by spermatozoa and converted to $\beta$-chlorolacetaldehyde. This inhibits the enzyme glyceraldehyde-3-phosphate dehydrogenase which renders the sperm unable to metabolize glucose. This chemical does decrease sperm motility and fertility but because the enzyme is not sperm specific, it is too toxic for use.

## Summary

- The most important hormone of the male reproductive system is testosterone, an example of the group of hormones called the androgens which are synthesized in the testes from cholesterol. Testosterone is also secreted by the adrenal cortex in both men and women. The secretion of testosterone from the testes is controlled by gonadotrophin releasing hormone from the hypothalamus and luteinizing hormone (LH) from the anterior pituitary. Like other steroid hormones, androgens produce their effects by interaction with intracellular receptors and alteration of protein synthesis.

- Androgens are responsible for the development and maintenance of male physique. In the foetus, androgens induce the development of male genitalia irrespective of chromosomal gender. They are also responsible for the changes that occur at puberty, thus they control the development of the male secondary sexual characteristics: increased muscle mass, elongation of the long bones of the limbs, male pattern pubic hair growth, facial hair, increased aggression and increased libido. In combination with follicle stimulating hormone (FSH) from the anterior pituitary, testosterone also controls the production of spermatozoa that is initiated at puberty.

- A deficiency of androgens in the foetus results in failure to develop male primary, sexual characteristics. A deficiency in later childhood results in either a failure to undergo the normal changes of puberty, or a delayed puberty. Failure to secrete androgens in adult life results in infertility and regression of male secondary sexual characteristics. All of these conditions can be treated successfully by male hormone replacement therapy. Excessive secretion of androgens may result in precocious puberty, culminating in premature closure of the epiphyses of the long bones. Paradoxically, excessive androgen concentrations result in decreased sperm production and testicular atrophy due to inhibition of LH and FSH secretion by negative feedback.

- Androgenic hormones may be used therapeutically for their anabolic effect to increase body mass in persons suffering from wasting disorders; they may also be used to induce temporary infertility (contraception). Androgens have also been used illegally to enhance performance in sportsmen and women.

## Selected reading

Hadley, M.E., 1992, Hormones and male reproductive physiology. In Hadley, M.E., *Endocrinology*, 3rd edn, Englewood Cliffs: Prentice-Hall, 455–475

Laycock, J. and Wise, P., 1996, Male reproductive endocrinology. In Laycock, J. and Wise, P., *Essential Endocrinology*, 3rd edn, Oxford: Oxford University Press, 145–162

Winters, S.J., 1994, Androgens and antiandrogens. In Brody, T.M., Larner, J. and Neu, H.C. (eds), *Human Pharmacology: Molecular to Clinical*, 2nd edn, St Louis: Mosby, 501–504

Zajac, J.D. and Warne, G.L., 1995, Disorders of sexual development. In Thakker, R.V. (ed.), *Baillière's Clinical Endocrinology and Metabolism, International Practice and Research. Vol. 9, No. 3, Genetic and Molecular Biological Aspects of Endocrine Disease*, London: Baillière Tindall, 555–579

# 11 The Ovaries and the Female Reproductive System

## 11.1 Introduction

The most important hormones of the female reproductive system are the oestrogens and progesterone. The oestrogens are a group of hormones, originally defined by their ability to induce a period of sexual receptivity (oestrus) in animals, the most important of which in humans are $17\beta$-oestradiol, oestrone and oestriol. They are synthesized and secreted by the ovaries under the control of the pituitary gonadotrophin, follicle stimulating hormone. Progesterone is the predominant member of a group of hormones called the progestogens. In females the primary source of progesterone is the ovaries although during pregnancy the foetoplacental unit assumes that role. In both males and females small amounts of progesterone are secreted by the adrenal cortex. The synthesis of progesterone by the ovaries is controlled by luteinizing hormone (see Figure 11.1).

## 11.2 Synthesis and mechanism of action of female sex hormones

As shown in Figure 11.2, oestrogens are produced in the ovaries by the conversion of cholesterol. The initial steps of this conversion involve synthesis of pregnenolone which is then converted to androstenedione or testosterone, in some cases via progesterone, in a manner similar to that seen in the testes and the adrenal cortex. The testosterone is then converted to $17\beta$-oestradiol by the aromatase enzymes, with a small amount of oestrone being produced from the androstenedione. Once secreted oestone can be converted in the periphery to either $17\beta$-oestradiol or oestriol. The most potent of the secreted oestrogens is $17\beta$-oestradiol, sometimes known as E2, followed by oestrone (E1) and finally oestriol (E3). Adipose tissue also expresses aromatase enzymes and is therefore able to convert androgens to oestrogens, most commonly androstenedione to oestrone. In males and postmenopausal females adipose tissue is a major source of oestrogens.

Once secreted the oestrogens are transported within the plasma bound to plasma proteins; approximately 70 per cent of the oestrogens are bound to sex hormone binding globulin (SHBG) with a further 25 per cent being bound to albumin. The biological actions of the oestrogens are produced by their interaction with a specific

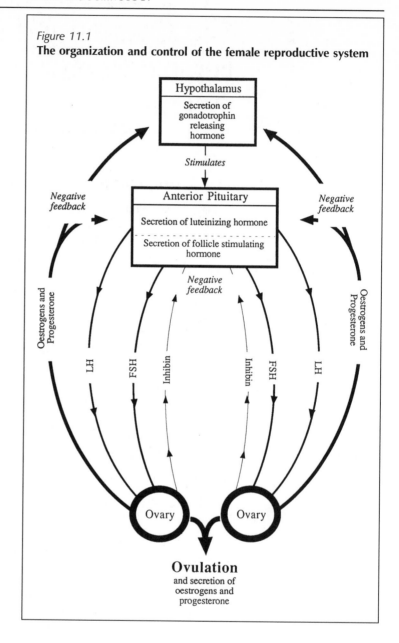

Figure 11.1

**The organization and control of the female reproductive system**

oestrogen receptor found within the cytoplasm and nucleus of target cells. There is evidence that the receptor normally 'shuttles' between these two sites dependent upon the hormonal environment. The effect of the hormone–receptor interaction is an alteration in gene expression, and therefore an alteration in protein synthesis (see Chapter 1). Degradation of the oestrogens occurs in the liver where they are converted to more water-soluble conjugates by sulphation and glucuronidation prior to excretion in the urine or the bile.

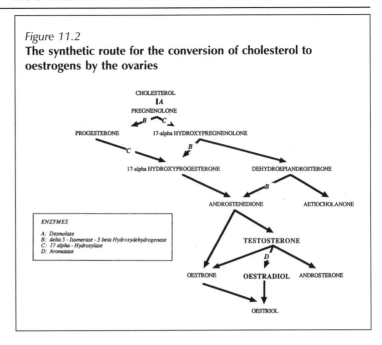

Figure 11.2
**The synthetic route for the conversion of cholesterol to oestrogens by the ovaries**

ENZYMES

A: Desmolase
B: delta 5 - Isomerase - 3 beta Hydroxydehydrogenase
C: 17 alpha - Hydroxylase
D: Aromatase

Progesterone is synthesized from pregnenolone as an intermediate during the production of oestrogens by the ovaries, and androgens by the testes and adrenal cortex, but it is also an important hormone in its own right; the major sources of progesterone are the ovaries and the foetoplacental unit. Once secreted, progesterone is transported in the plasma bound to corticosteroid binding globulin (CBG) and albumin; only about 2 per cent remains unbound. Progesterone is rapidly degraded in the liver, thus its plasma half-life is approximately five minutes.

The mechanism of action of progesterone is similar to that of the other steroid hormones in that it acts on specific progesterone receptors found in the cytoplasm and nucleus to influence gene expression and protein synthesis. It is important to note that progesterone receptors are normally synthesized as a result of oestrogen activity at the oestrogen receptor, thus tissues normally require 'priming' by oestrogens before they are able to respond to progesterone. There is now also good evidence that progesterone is able to produce an effect on cell membranes; this action is mediated by an effect of the hormone on membrane bound receptors for the inhibitory neurotransmitter $\gamma$-aminobutyric acid (GABA).

It is interesting to note that the circulating concentrations of progesterone and testosterone in both males and females are greater than the concentrations of $17\beta$-oestradiol. In males, plasma progesterone concentrations are approximately 1 nmol/l (0.3 ng/ml) while testosterone concentrations are 10–30 nmol/l (3–9 ng/ml); concentrations of $17\beta$-oestradiol are below 0.2 nmol/l (0.05 ng/ml). In females plasma concentrations of $17\beta$-oestradiol vary between 0.08 and 1.8 nmol/l (0.02–0.5 ng/ml) with plasma progesterone concentrations varying between 5 and 50 nmol/l (1.5–15 ng/ml).

Circulating concentrations of testosterone and progesterone are greater than those of oestradiol in both males and females.

Testosterone concentrations in females, at 0.5–2.5 nmol/l (0.2–0.7 ng/ml), are comparable with the plasma concentration of $17\beta$-oestradiol. These relative concentrations demonstrate the relative potency of oestrogenic hormones.

## 11.3 Development of female primary sexual characteristics

As described in detail in Chapter 10, the gender of a developing foetus is dependent upon its genetic make-up, thus possession of two X chromosomes normally results in the development of female primary sexual characteristics while the possession of one X and one Y chromosome normally results in the development of male primary sexual characteristics. In the early stages of development the embryo possesses gonads and associated structures, the Müllerian and Wolffian ducts, which are capable of developing into either male or female internal reproductive organs. In the absence of the Y chromosome and the Sry gene the gonads develop as ovaries, the Wolffian ducts atrophy and the Müllerian ducts differentiate to form the Fallopian tubes, uterus and vagina (see Chapter 10, Figure 10.2). The embryonic/foetal ovaries do not synthesize or secrete oestrogens or progesterone.

> The foetal ovaries secrete neither oestrogens nor progesterone.

In the absence of stimulation by testosterone the embryonic genitalia develop with a clitoris and labia. It can therefore be seen that in the absence of the Y chromosome and testosterone the embryo will develop female internal and external genitalia; female sex hormones have no effect of the embryo/foetus and are therefore not required for the development of a female anatomy. Biologically, this lack of effect of oestrogens or progesterone on the embryo is important because it removes any potential influence of the high circulating concentrations of the maternal female sex hormones.

> The high concentrations of circulating maternal female sex hormones during pregnancy have no effect on the foetus.

## 11.4 Ovarian structure and oogenesis

The function of the ovary is the production of mature gametes (ova) and the synthesis and secretion of the female sex hormones. The ability to produce mature ova (oogenesis) is established very early in foetal development, but the ovaries do not begin to produce the sex hormones until puberty. The ovaries are situated within the abdomen, one lying either side of the uterus. Each one consists of three regions: the hilum, which is the point of attachment to the surrounding structures, and through which the blood vessels, nerves and lymphatics run; the outer cortex; and the inner medulla. During embryogenesis approximately 3 million oocytes, each with the potential to develop into a viable ovum, are formed within the cortex of each ovary. Many primary follicles become arrested at this stage and never develop further, while after puberty a small number develop further to secondary and Graafian follicles (see later). By the time of birth the number of primary follicles has been reduced to

Females are born with approximately 1 million ova, but this is reduced to 250 000 by the time of puberty.

approximately 1 million by a process called atresia; atresia continues throughout life such that there may be only 250 000 oocytes remaining at puberty. The development, maturation and release (ovulation) of ova occurs after puberty in a cyclical manner controlled by the hormones of the anterior pituitary gland (see later).

## 11.5 Effects of the female sex hormones and the development of secondary sexual characteristics

The development of female secondary sexual characteristics is induced by an increase in the secretion of oestrogens at puberty. Progesterone has little effect other than to promote and maintain pregnancy (see later). Prior to puberty the secretion of follicle stimulating hormone (FSH) and luteinizing hormone (LH) is negligible, although as puberty approaches the circulating concentrations begin to rise, on to which is superimposed increasingly frequent pulses of secretion. There is negligible secretion of ovarian steroids prior to puberty, but there is a slight rise in the secretion of oestrogens as puberty approaches; the adrenal cortex also begins to secrete the androgen dehydroepiandrostenedione approximately two years before the onset of the pubertal changes. The stimulus that causes the onset of puberty is unknown but it is believed to be a function of the rate of maturation of the hypothalamus and may be related to a decreasing sensitivity of the hypothalamus to negative feedback. Prior to puberty the secretion of androgens by the adrenal cortex may be sufficient to inhibit secretion of LH and FSH by the anterior pituitary gland. The onset of puberty is heralded by the decreased sensitivity to negative feedback and therefore increased secretion of LH, FSH and ovarian steroids.

The clearest sign of puberty in females is the initiation of menstruation.

The first sign of puberty in girls is the initiation of breast development, closely followed by the appearance of pubic hair along the labia. Fifty per cent of girls have developed breast buds by the age of 11 years 3 months and pubic hair by the age of 11 years 6 months; development of adult breasts and pubic hair pattern does not occur until after the age of 15. The initiation of breast development, referred to as telarche, is dependent upon the presence of oestrogens while the growth of pubic hair is probably stimulated by the increased secretion of androgens by the adrenal cortex, the adrenarche. The pattern of pubic hair growth in females is different from that in males indicating that there is some influence of oestrogens. The clearest sign of puberty is the initiation of menstruation, the menarche; this occurs in 50 per cent of girls by the age of 13. Menstruation, and the events of the menstrual cycle, are controlled by the pulsatile secretion of FSH and LH by the anterior pituitary gland. The details of the menstrual cycle are covered in the following section.

As well as breast development and the development and maintenance of the internal and external female genitalia, oestrogens, in

In animals, female sexual activity is increased by the administration of oestrogens.

conjunction with growth hormone, also cause a spurt of long bone growth at the time of puberty (see Figure 11.3). In girls this growth spurt normally begins about two years earlier than it does in boys, at about 12 years of age. Cessation of growth in girls occurs at about 16 compared with about 18 in boys. Prior to puberty there is little difference in height between the sexes, but because the growth spurt in boys occurs later, from a greater starting point, boys are generally approximately 10 cm taller than girls by the age of 18. Oestrogens are also responsible for decreased sebaceous gland activity in females, giving rise to a fairer complexion and the typical pattern of subcutaneous fat deposition seen in females who have a more rounded appearance compared to the angular, muscular appearance of males. Oestrogens also have psychological effects leading to a decrease in aggression. There is some dispute about the effects of oestrogens on libido; in many animals, female sexual activity is increased by the administration of oestrogens; in humans, however, sexual activity is unaffected by the decreased secretion of oestrogens that occurs at the menopause. Indeed, in some women decreased libido is associated with decreased secretion of androgens and is treated by androgen supplements. In humans it appears that androgens are the major hormonal determining factor in the control of female sexual activity.

## 11.6 The menstrual cycle

As illustrated in Figure 11.4, the menstrual cycle is characterized by cyclical changes in the female internal genitalia, brought about by the cyclic variations in oestrogen and progesterone secretion. These changes in oestrogen and progesterone secretion are themselves

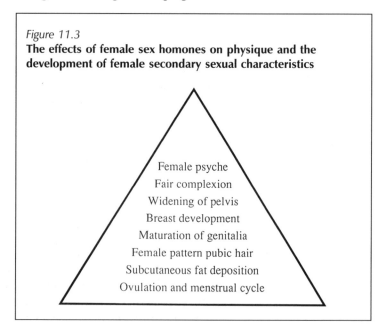

*Figure 11.3*
**The effects of female sex homones on physique and the development of female secondary sexual characteristics**

Female psyche
Fair complexion
Widening of pelvis
Breast development
Maturation of genitalia
Female pattern pubic hair
Subcutaneous fat deposition
Ovulation and menstrual cycle

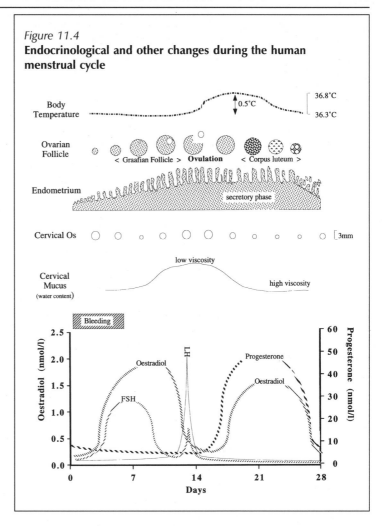

Figure 11.4

**Endocrinological and other changes during the human menstrual cycle**

controlled by LH and FSH from the anterior pituitary gland and GnRH from the hypothalamus.

The menstrual cycle begins with the secretion of FSH from the anterior pituitary, under the influence of GnRH. The effect of this FSH is to stimulate the maturation of 6 to 12 primary follicles within the ovary to become secondary follicles. Under the influence of FSH the oocyte increases in size and the cells surrounding it begin to replicate and to secrete a mucoid material to form the zona pellucida. Of these 6–12 secondary follicles, only one develops further; the remainder undergo atresia. In the one remaining secondary follicle the cells surrounding the oocyte continue to replicate while adjacent interstitial cells of the ovary become the theca interna and the theca externa. In the presence of low levels of LH the thecal cells begin to synthesize androgens, predominantly androstenedione, from cholesterol. This effect of LH is analogous to its effect on the Leydig cells of the testes in that it regulates the rate-limiting conversion of cholesterol to pregnenolone. Meanwhile, the cells immediately surrounding the oocyte produce a peptide called inhibin and, under the influence of FSH, produce aromatase enzymes capable of converting the androgens to $17\beta$-oestradiol. The secreted oestrogens, together with inhibin, have the effect of reducing the secretion of FSH; they also have a local effect of stimulating further development of the follicle, independent of FSH, so that a large vacuole appears in the centre. At this stage the fully developed mature follicle is known as the Graafian follicle.

Under the influence of a sudden surge of LH secretion, accompanied by a lesser peak of FSH secretion, the follicle ruptures and the ovum is expelled into the peritoneal cavity. Also under the influence of LH, the remnants of the follicle, comprised of the epithelial (granulosa) cells which formerly surrounded the oocyte together with a small number of thecal cells, accumulates large quantities of cholesterol and becomes the corpus luteum. This corpus luteum is capable of synthesizing and secreting large amounts of progesterone, together with smaller amounts of oestrogens which suppress the secretion of LH and FSH. In the absence of these gonadotrophins the lifespan of the corpus luteum is approximately seven days after which the secretion of progesterone and $17\beta$-oestradiol decreases and the cells atrophy.

As a result of the decreased secretion of progesterone and $17\beta$-oestradiol by the corpus luteum the anterior pituitary gland is freed from negative feedback and gonadotrophin secretion resumes, resulting in the development of another 6–12 primary follicles and the initiation of another cycle.

The patterns of hormones secretion throughout the normal menstrual cycle are shown in Figure 11.4. These variations in hormone secretion also bring about changes in the physiology and anatomy of the reproductive organs and related structures. As described above, the initial secretion of FSH causes development and maturation of the ovarian follicle during the early (follicular) phase of the menstrual cycle; this is accompanied by synthesis and secretion of $17\beta$-oestradiol by that follicle. As a result of the increase in oestrogen secretion the endometrium enters the proliferative phase in which it begins to thicken, with a rich blood supply; and the cervix begins to dilate and secrete a thin, watery mucus. By the time of the LH surge on day 13 of the cycle the cervix is fully dilated and the cervical mucus is at its least viscous so that sperm transport to the uterus is relatively unhindered. Ovulation occurs approximately 24 hours after the LH surge, after which the ovum travels down the Fallopian tubes towards the uterus and the remaining cells of the follicle become the corpus luteum. The corpus luteum secretes progesterone and smaller amounts of $17\beta$-oestradiol. Under the influence of progesterone the endometrium continues to thicken and to secrete glycoproteins which offer nutritional support to the ovum, should it become fertilized; this is called the secretory phase. The cervix also begins to constrict and the cervical mucus becomes more viscous during this phase. The secreted progesterone also has effects away from the reproductive organs: progesterone acts on the hypothalamus to cause an increase in body temperature of

0.5–1.0°C and, in some women, acts on the breast tissue to cause a transient enlargement of the breasts throughout the latter stages of the menstrual cycle.

The corpus luteum has a lifespan of about seven days, after which the secretion of progesterone and 17$\beta$-oestradiol decreases, returning to basal levels by about day 28 of the cycle. As a consequence of the declining steroid secretion the thickened, secretory endometrium can no longer be maintained and it therefore begins to undergo necrosis. The initial phase of the necrotic process is temporary constriction of the blood vessels within the endometrium; the secretory tissues are therefore deprived of oxygen. As the endometrial cells die they release a group of hormones known as prostaglandins into the local environment. Prostaglandins are members of a group of chemicals known as 'pain producing substances' and they can also cause uterine contraction. The blood vessels then dilate and there is an increase in blood flow to the endometrium, which is now dead, which has the effect of dislodging the cell debris from the endometrial surface. In the first instance the blood clots, but this is followed by fibrinolysis of the clot so that the menstrual fluid that is lost consists of fibrinolysed blood together with endometrial cell debris. Menstruation begins on about the twenty-eighth day of the menstrual cycle. The prostaglandins that are released at this time are responsible for the menstrual cramps and 'period pain' that might be experienced; thus drugs such as aspirin and paracetamol, which inhibit prostaglandin synthesis, are of use in the treatment of dysmenorrhoea. Some workers believe, however, that paracetamol is the drug of choice because the inhibitory effect of aspirin on blood clotting could increase menstrual blood loss. In some women, the hormonal changes that occur during the days preceding the onset of menstruation are also believed to produce psychological changes (see Box 11.1). By convention, the first day of menstruation is designated as day 1 of the next menstrual cycle.

## 11.7 The physiology of pregnancy, parturition and the puerperium

Deposition of spermatozoa into the vagina, and their subsequent progression through the cervical canal and into the uterus may result in fertilization of the released ovum. An ovum remains capable of being fertilized for 24–36 hours following ovulation, while the spermatozoa remain viable for 48–72 hours after deposition. It has been shown that the probability of fertilization is approximately 60 per cent when coitus occurs immediately prior to or on the day of ovulation. Fertilization normally occurs while the ovum is passing through the Fallopian tube.

The fertilized ovum continues its passage through the Fallopian tube and reaches the uterus three or four days after ovulation. Within the uterus the fertilized ovum, now termed a blastocyst, continues to undergo cell division, obtaining nutrients and oxygen

*Box 11.1*   **Premenstrual syndrome**

Some women experience a range of psychological and physical symptoms such as depression, breast tenderness and sleep disorders during the premenstrual period. If the symptoms occur consistently at this time, and resolve rapidly with the onset of menstruation they are often labelled as premenstrual syndrome (PMS). It is estimated that 50 per cent of women suffer from PMS at some time during their reproductive life and that 10 per cent of women suffer to a serious extent. It is commonly believed that the cause of PMS is an 'imbalance' of the reproductive hormones, however, despite extensive research, no differences in oestrogens, progesterone, aldosterone or any other hormone has ever been demonstrated between sufferers and non-sufferers. It should be noted, however, that many of the early studies used undergraduates or nurses aged 18–25 as their subjects even though PMS frequently does not onset until the late twenties or mid-thirties. More recent studies have investigated the older age group, but there are still no demonstrable links between hormones and severity or incidence of PMS. Despite these negative results, many PMS sufferers have been treated with progestogens with some apparent success, although the efficacy of the treatment has never been proven in controlled clinical trials. Many other treatments for PMS exist, but again there are few, if any, controlled studies which demonstrate efficacy. These studies are hampered to some extent by the fact that placebo has been shown to provide temporary relief (up to three months) in 30–45 per cent of cases, which demonstrates the unpredictable nature of the condition.

from the endometrial secretions. After about six days within the uterus the blastocyst fuses with cells of the endometrium; this is implantation. There is considerable ethical debate as to whether pregnancy begins with fertilization or with implantation. This is particularly important when considering the actions of some forms of contraception (see later). The effect of implantation is to cause the development of endometrial capilliaries which grow towards the blastocyst. It can be shown that the presence of oestrogens during the luteal phase is essential for implantation to occur. The changes that take place in the endometrium following implantation are termed 'decidualization'. At the same time, the trophoblast begins to develop a vascular system which grows towards the maternal vessels to form the placenta. The placenta allows exchange of gaseous and metabolic products between the foetal and maternal circulation without there being any mixing of blood; it also acts as a barrier which can offer selective transport of various moieties such as proteins and cells.

The presence of an implanted trophoblast can be detected by its synthesis of human chorionic gonadotrophin (hCG), a glycoprotein with a structure similar to that of LH. The function of hCG is to prolong the activity of the corpus luteum, which would, by now, normally be beginning to atrophy because of the declining secretion of LH from the maternal anterior pituitary gland. Secretion of hCG increases rapidly from the point of implantation to about 30–40 days after conception after which it begins to decline. The main function of the hCG is to increase the secretion of progesterone and oestrogens by the corpus luteum until about the eighth week of pregnancy when the foetoplacental unit begins to synthesize its

The presence of hCG in the urine has allowed the development of cheap pregnancy tests for use in the home.

own steroid hormones. It is also probable that hCG has the function of stimulating the secretion of oestrogens, progesterone and, in the case of a male foetus, androgens, by the foetoplacental unit throughout pregnancy. The detection of hCG in the maternal urine forms the basis of the commercially available pregnancy test kits. The ability to detect urinary hCG at concentrations as low as 25 IU/l means that pregnancy can now be detected before the first missed period, although not all of the detected pregnancies culminate in successful, full-term pregnancies (see Box 11.2).

The corpus luteum also secretes a peptide hormone called relaxin which inhibits uterine motility, softens the cervix at the time of delivery and relaxes the ligaments of the synthesis pubis, although the importance of the hormone for the progression of a normal pregnancy is debatable.

Under the influence of hCG, maternal plasma concentrations of progesterone rise steadily throughout pregnancy reaching levels of approximately 500 nmol/l, which are ten times greater than the peak plasma progesterone concentration during the menstrual cycle. Apart from during the early stages of pregnancy, the principal source of progesterone secretion is the placenta. Progesterone is freely able to cross the placenta. The plasma oestrogen

*Box 11.2* **Pregnancy testing**

The traditional indications of pregnancy are amenorrhoea, morning sickness and weight gain; however there is an increasing need to confirm pregnancy before these signs become apparent, in order to adapt one's lifestyle. This demand lead to the development of home-use pregnancy tests. All pregnancy tests act by detecting the presence of human chorionic gonadotrophin (hCG) in either the blood or the urine. hCG is a peptide hormone which is produced by the foetoplacental unit and acts to prolong the life-span of the corpus luteum; it is chemically related to LH, FSH and TSH. Early pregnancy tests were unable to differentiate hCG from the pituitary hormones, thus they would only indicate a positive result when the concentrations of hCG were much higher than normal physiological concentrations of either LH, FSH or TSH. This meant that many of the tests would not give a positive pregnancy test result until about the third month of pregnancy. The newer pregnancy tests are able to measure hCG specifically, without interference from the other hormones; thus they are now able to detect raised hCG (pregnancy) even before the first 'missed period'.

The introduction of such sensitive tests has, however, brought new problems. Since their introduction there has been an increase in the rate of miscarriage and spontaneous abortion. This is not to suggest that the pregnancy tests cause these abortions, but because up to 60 per cent of conceptions normally fail before the third month of pregnancy, previously these would have been labelled as 'a late period'; the advent of the new tests means that they are now confirmed as a failed pregnancy.

Other possible problems with the tests are that they do not differentiate the source of the hCG – a positive test result can be obtained in the presence of an hCG secreting tumour; some forms of treatment of infertility involve the administration of hCG (proprietary names: Gonadotrophon, Pregnyl, Profasi) which is excreted in the urine and therefore guaranteed to give a positive pregnancy test result; and a positive test result may be obtained for up to two weeks after a miscarriage, termination of pregnancy or parturition.

It should therefore be remembered that a positive pregnancy test result indicates the presence of hCG in the urine rather than the impending arrival of a baby, and that there was wisdom in the old tradition of waiting until the 'third missed period' before announcing a pregnancy.

concentrations also rise greatly during pregnancy. Oestrogens are derived by the placenta from the precursor dehydroepiandrostenedione sulphate which is produced by the maternal and foetal adrenal glands. Several oestrogens are secreted, but the predominant oestrogen of pregnancy is oestriol which can only be derived from foetal precursors. By the end of pregnancy circulating oestrogens are approximately 300 times greater than the peak menstrual cycle values. Determination of oestriol concentrations allows the health of the foetus to be monitored.

Another placental hormone is human placental lactogen (HPL) which is secreted in increasing amounts throughout pregnancy. HPL has a chemical structure similar to those of prolactin and growth hormone. The actions of HPL are unclear but it causes some breast development during pregnancy and has an effect similar to that of hCG; HPL also causes lipolysis. Like oestriol, the unique foetal origin of HPL mean that it can be used to monitor foetal well-being.

The demands of the foetus upon the mother also cause increased secretion of many other hormones during pregnancy. There is an increase in secretion of thyroid hormones, adrenal corticosteroids and parathyroid hormone, although in the case of the former two the effects of the increased hormone secretion are partially attenuated by an increase in plasma protein binding.

The process of pregnancy culminates in parturition during which the foetus and placenta are expelled from the uterus by contractions of the myometrium. Throughout pregnancy the muscle fibres of the uterus enlarge and become connected by gap junctions which allow easy, rapid spread of muscle action potentials. The presence of the gap junctions mean that the uterus acts as a functional syncytium with coordinated muscle contraction. During pregnancy the increased spontaneous activity of the oestrogen dominated uterus is inhibited by the presence of progesterone. These hormone effects are mediated partially by the oestrogen-stimulated and progesterone-inhibited synthesis of oxytocin receptors and prostaglandins. It is possible that the stimulus for the initiation of parturition is a decrease in the relative progesterone–oestrogen ratio. Several studies have reported a drop in plasma progesterone concentration at about 38 weeks of pregnancy, although not all women exhibit such a pattern of secretion. Some studies suggest that the change in the progesterone–oestrogen ratio may result from an increased activity of the foetal adrenal glands resulting in increased conversion of progesterone to androgens, and thence oestrogens, and that the changes need only be local to the uterus for parturition to commence. Oxytocin and prostaglandins are also probably important for the process of parturition (see Chapter 3).

Following delivery of the foetus and the placenta the maternal source of oestrogens and progesterone is removed; this leads to a precipitous fall in circulating concentrations. Within the first few days postpartum (the puerperium) the decline in steroid activity results in a reduction in the renal sodium reabsorption that was stimulated during pregnancy by the aldosterone-like activity of the sex steroids. As a result, daily urine excretion may increase from a normal, non-pregnant value of 500–700 ml to a value exceeding 3000 ml. The sudden fall in hormone concentrations may be responsible for the emotional disturbances that are seen in approximately 50 per cent of mothers ('baby blues'), although no relationship between puerperal mood and hormones has ever been demonstrated. The decline in circulating oestrogens also allows the initiation of lactation (Chapter 2).

## 11.8 Physiology of the menopause and hormone replacement therapy (HRT)

The average age of menopause in the UK is 51 and in Central Africa is 44.

The menstrual cycle continues throughout the reproductive life, normally ceasing only during pregnancy. The cessation of menstruation in later life is called the menopause, but the menstrual cycle may be very irregular in the years preceding the menopause. The cause of the menopause is unknown, but the most important feature is a failure of the development of ovarian follicles and failure of secretion of oestrogens and progesterone; this may reflect depletion of the primary follicles. Some workers have suggested that the onset of the menopause may be a hypothalamic event, with a loss of the rhythmic secretion of GnRH; however recent studies indicate that the rhythmic, pulsatile secretion of LH and FSH continues after the menopause. This finding strengthens the belief that the menopause is a result of ovarian failure. The average age of menopause in the UK is 51, while in Central Africa it is approximately 44; in the UK menopause prior to the age of 40 is considered abnormal. There is evidence that the age of menopause is increasing, probably because of improving nutritional standards.

The first sign of the menopause is the increasingly irregular nature of the menstrual cycle. Nearer to the actual date of the menopause women may suffer episodic sensations of heat in the head, neck and upper chest; in the UK these symptoms are usually labelled as 'hot flushes' while in the USA they are termed 'hot flashes'. These may be due to alterations of GnRH secretion, although $\alpha$-adrenoceptors are also involved. Hot flushes subside after the cessation of menstruation. The actual cessation of menstruation is termed the menopause, while the years either side of that event are termed the climacteric.

The menopause results in a decrease in the secretion of the female sex hormones, and there is therefore the possibility of some regression of the female secondary sexual characteristics, for example atrophy of the endometrium and atrophic vaginitis. The most serious sequelae of the menopause are the loss of the protein matrix of the bone (osteoporosis) and cardiovascular disease.

Postmenopausal women may lose between 1 and 3 per cent of their bone mass each year with a resultant tenfold increase in the risk of fractures. The loss of oestrogens also removes their 'protective' effect against the thrombogenic effects of the endogenous androgens, thus the incidence of coronary thrombosis in postmenopausal women is similar to that in males of the same age. The menopause is also accompanied by an increased incidence of psychiatric disorders, especially depression, although no definite links have been demonstrated between the mood disorders and hormone concentrations.

By administration of oestrogens, either transdermally or orally (see Table 11.1), the hot flushes can be controlled, the regression of secondary sexual characteristics can be inhibited, the onset of osteoporosis can be delayed and the risk of cardiovascular disease reduced by the reduction of plasma cholesterol and low density lipoprotein cholesterol. Oestrogen therapy, however does carry risks. Oestrogen may promote existing breast cancer (see later) and may promote the proliferation of the endometrium or endometrial carcinoma; oestrogen replacement therapy increases the risk of endometrial carcinoma three-fold.

Oestrogen only replacement therapy, therefore, is only appropriate for women who have undergone hysterectomy. In women in whom the uterus remains, the risks of endometrial hyperplasia can be decreased by cyclical administration of a progestogen. The administration of a progestogen, followed by its sudden withdrawal, causes a menstrual-like bleed which reduces the risk of the oestrogen-induced endometrial hyperplasia. There is therefore a regular 'period', although this is seen as undesirable by some women. In some cases the progestogen can be given continuously, but irregular bleeding may still ensue.

Some of the orally active progestogens used for the prevention of endometrial hyperplasia possess androgenic properties and therefore produce adverse effects on blood lipids. In these cases the coadministration of the progestogen partially negates the beneficial effects of the oestrogens on the low density and high density lipoprotein cholesterol. The epidemiological data suggest that such combined treatment still offers protection against cardiovascular disease, but not to the same extent as that offered by the use of unopposed oestrogens, and that the cyclical progestogen reduces the risk of endometrial carcinoma to that of an untreated population.

*Use of HRT for up to ten years does not increase the risk of breast cancer.*

Concern over the risks of breast cancer is one of the major reasons for discontinuing hormone replacement therapy. There is considerable evidence that oestrogens are involved in the aetiology of breast cancer, and therefore administration of additional oestrogens postmenopausally might be expected to increase the risk of developing breast cancer. The available evidence suggests that use of oestrogen replacement therapy for less than ten years has no significant effect on the number of breast cancer cases, but that the relative risk of developing the disease is increased by 50 per cent after 20 years of therapy. Progestogens have no effect on the incidence of breast cancer.

*The cost of routine provision of HRT is less than that of the treatment of the consequences of the menopause.*

It is generally believed that use of combined oestrogen/progestogen hormone replacement therapy for ten years is sufficient to remove the risk of osteoporosis without inducing serious side-effects; five years of treatment reduces the incidence of osteoporosis

**Table 11.1** Hormonal preparations available for use in postmenopausal hormone replacement therapy (reproduced, with permission of the publisher, from the *Monthly Index of Medical Specialities (MIMS)*, in which details of available preparations are updated monthly)

## Hormone Replacement Therapy

| Type | Brand | Oestrogen | Progestogen | Formulation | Strengths (of oestrogen) | Bleed | RX* Charge | Cost/ 28 days |
|---|---|---|---|---|---|---|---|---|
| **SYSTEMIC** |||||||||
| Sequential combined therapy | Climagest | Oestradiol | Norethisterone | Tabs | (1mg, 2mg) | M | 2 | £4.38 |
| | Cyclo-progynova | Oestradiol | Levo/norgestrel | Tabs | (1mg, 2mg) | M | 2 | £3.50 |
| | Elleste Duet | Oestradiol | Norethisterone | Tabs | (1mg, 2mg) | M | 2 | £3.24 |
| | Estracombi | Oestradiol | Norethisterone | Patches | (50mcg) | M | 2 | £11.14 |
| | Estrapak | Oestradiol | Norethisterone | Patches + Tabs | (50mcg) | M | 2 | £9.48 |
| | Evorel-Pak | Oestradiol | Norethisterone | Patches + Tabs | (50mcg) | M | 2 | £8.45 |
| | Evorel Sequi | Oestradiol | Norethisterone | Patches | (50mcg) | M | 2 | £11.00 |
| | Femapak | Oestradiol | Dydrogesterone | Patches + Tabs | (40mcg, 80mcg) | M | 2 | £8.45 £8.95 |
| | Femoston | Oestradiol | Dydrogesterone | Tabs | (1mg,2mg) | M | 2 | £4.99 |
| | Femoston 2/20 | Oestradiol | Dydrogesterone | Tabs | (2mg) | M | 2 | £7.48 |
| | Improvera | Oestrone | Medroxyprogesterone | Tabs | (0.93mg) | M | 2 | £3.95 |
| | Menophase | Mestranol | Norethisterone | Tabs | 1 strength | M | 2 | £3.25 |
| | Nuvelle | Oestradiol | Levonorgestrel | Tabs | (2mg) | M | 2 | £4.59 |
| | Nuvelle TS | Oestradiol | Levonorgestrel | Patches | (80mcg, 50mcg) | M | 2 | £11.00 |
| | Premique Cycle | Conj. oestrogens | Medroxyprogesterone | Tabs | (0.625mg) | M | 2 | £7.54 |
| | Prempak-C | Conj. oestrogens | Norgestrel | Tabs | (0.625, 1.25mg) | M | 2 | £4.46 |
| | Tridestra | Oestradiol | Medroxyprogesterone | Tabs | (2mg) | Q | 2 | £8.30 |
| | Trisequens | Oestradiol/ oestriol | Norethisterone | Tabs | 2 strengths | M | 2 | £6.85 |
| Continuous combined therapy | Climesse | Oestradiol | Norethisterone | Tabs | (2mg) | X | 1 | £7.90 |
| | Evorel Conti | Oestradiol | Norethisterone | Patches | (50mcg) | X | 1 | £12.90 |
| | Kliofem | Oestradiol | Norethisterone | Tabs | (2mg) | X | 1 | £8.65 |
| | Premique | Conj. oestrogens | Medroxyprogesterone | Tabs | (0.625mg) | X | 1 | £7.54 |
| Gonadomimetic | Livial | | | Tabs | 1 strength | X | 1 | £13.66 |
| Unopposed oestrogen | Climaval | Oestradiol | | Tabs | (1mg, 2mg) | | 1 | £2.34 |
| | Dermestril | Oestradiol | | Patches | (25, 50, 100mcg) | | 1 | £5.75, £6.35, £6.99 |
| | Elleste Solo | Oestradiol | | Tabs | (1mg, 2mg) | | 1 | £1.78 |
| | Elleste Solo MX | Oestradiol | | Patches | (40, 80mcg) | | 1 | £5.96 £6.56 |
| | Estraderm TTS/ Estraderm MX | Oestradiol | | Patches | (25, 50, 100mcg) | | 1 | £6.75, £7.45, £8.20 |
| | Evorel | Oestradiol | | Patches | (25, 50, 75, 100mcg) | | 1 | £6.75, £7.45, £7.90, £8.20 |
| | Fematrix | Oestradiol | | Patches | (40mcg, 80mcg) | | 1 | £6.45, £6.95 |
| | Femseven | Oestradiol | | Patches | (50, 75, 100 mcg) | | 1 | £6.44, £7.49, £8.19 |
| | Harmogen | Oestrone | | Tabs | (0.93mg) | | 1 | £3.14 |
| | Hormonin | Oestriol/ oestradiol/ oestrone | | Tabs | 1 strength | | 1 | £2.00 |
| | Menorest | Oestradiol | | Patches | (37.5, 50, 75mcg) | | 1 | £6.34, £6.44, £7.50 |
| | Oestrogel | Oestradiol | | Gel | (1.5mg) | | 1 | £7.95 |
| | Premarin | Conj. oestrogens | | Tabs | (0.625, 1.25mg, 2.5mg) | | 1 | £2.45, £3.33, £3.55 |
| | Progynova | Oestradiol | | Tabs | (1mg, 2mg) | | 1 | £2.34 |
| | Progynova TS | Oestradiol | | Patches | (50,100mcg) | | 1 | £6.44, £8.33 |
| | Sandrena | Oestradiol | | Gel | (0.5mg, 1mg) | | 1 | £5.95, £6.85 |
| | Zumenon | Oestradiol | | Tabs | (1mg, 2mg) | | 1 | £2.55 |
| Adjunctive progestogen | Crinone | | Progesterone | Vaginal Gel | 1 strength | | | £11.60 |
| | Duphaston HRT | | Dydrogesterone | Tabs | 1 strength | | 1 | £2.77 |
| | Micronor HRT | | Norethisterone | Tabs | 1 strength | | 1 | £1.25 |

**Table 11.1** Cont'd

| Hormone Replacement Therapy | | | | | | | | |
|---|---|---|---|---|---|---|---|---|
| Type | Brand | Oestrogen | Progestogen | Formulation | Strengths (of oestrogen) | Bleed | RX* Charge | Cost/ 28 days |
| **LOCAL** | | | | | | | | |
| Oestrogen only | Estring | Oestradiol | | Vaginal ring | | | 1 | |
| | Ortho Dienoestrol | Dienoestrol | | Vaginal cream | | | 1 | |
| | Ortho-Gynest | Oestriol | | Pessary Vaginal cream | | | 1 1 | |
| | Ovestin | Oestriol | | Vaginal cream | | | 1 | |
| | Premarin | Conj. oestrogens | | Vaginal cream | | | 1 | |
| | Tampovagan | Stilboestrol | | Pessary | | | 1 | |
| | Vagifem | Oestradiol | | Vaginal tabs | | | 1 | |

Bleed – M = Monthly; Q = Quarterly; X = No bleed          *Combination packs incur multiple prescription charges

by 50 per cent. There is still debate, however, about the cost effectiveness of HRT but its proponents suggest that the cost of routine provision of HRT for ten years would be more than recouped by the savings on orthopaedic surgery.

A recent advance in hormone replacement therapy has been the introduction of tibolone which possesses both oestrogenic and progestogenic activity. This drug can therefore be used without the need of additional hormone therapy, but its continuous administration means that there are no withdrawal, 'menstrual-like' bleeds.

## 11.9 Disorders of the female reproductive system and their treatment

Prior to puberty neither oestrogen nor progesterone plays any role in development or maturation, thus there are no clinical disorders associated with hormone deficiency. Early exposure to oestrogens, however, can cause premature development of secondary sexual characteristics. True precocious puberty is caused by pulsatile secretion of GnRH from a tumour, resulting in secretion of LH and FSH, and thus oestrogens and progestererone. In most cases the girls undergo the normal changes of puberty and experience menstruation. Treatment is usually by blockade of LH and FSH secretion using a GnRH analogue (Box 10.1, Chapter 10). In pseudoprecocious puberty there is development of secondary sexual characteristics due to inappropriate secretion of oestrogens from a tumour, or due to ingestion of exogenous oestrogens, but there is no initiation of menstrual cycles. As in males, both true and pseudoprecocious puberty in females are associated with an initial spurt in growth followed by early closure of the epiphyses. Ultimately these girls are shorter than their peers. Treatment is by removal of the source of the oestrogens.

Delayed puberty in females is manifested as primary amenorrhoea, where menstruation never occurs. Fifty per cent of girls menstruate by the age of 13, but 3 per cent of girls have not menstruated by the age of 15. The cause of primary amenorrhoea may lie in the hypothalamus or anterior pituitary gland or may be at the level of the ovaries. In Turner's syndrome, which occurs in approximately 3 per 10 000 female births, the absence of an X chromosome results in abnormal development of the ovaries. There is therefore a deficiency of oestrogen and progesterone synthesis and a raised secretion of LH and FSH due to reduced negative feedback. Such patients can be treated by oestrogen replacement therapy coupled with cyclical progestogens to induce a regular menstrual bleed. These women are normally infertile, although egg donation and *in vitro* fertilization has recently been used to initiate pregnancy. In cases of primary amenorrhoea due to hypothalamic or pituitary disorders menarche may sometimes be induced by administration of a GnRH analogue or an oestrogen receptor antagonist. Oestrogen receptor antagonists such as tamoxifen and clomiphene act by preventing the negative feedback effects and therefore increase the secretion of LH and FSH.

Secondary amenorrhoea is the cessation of normal menstrual cycles. The most common causes are pregnancy and menopause, although factors such as malnutrition and chronic stress may also induce amenorrhoea, as can hyperprolactinaemia, hyperthyroidism and any hypothalamic or pituitary gland disorder. Treatment is by resolution of the underlying disorder, sometimes in combination with clomiphene or tamoxifen therapy. Up to 10 per cent of cases of secondary amenorrhoea are caused by polycystic ovarian disease (Stein–Leventhal syndrome). In this condition there is increased secretion of androgens by the ovaries and the adrenal glands. The raised androgen secretion causes hirsutism and suppresses ovulation. Treatments include oestrogen administration, which suppresses LH secretion and therefore androgen secretion, and the use of androgen receptor antagonists.

> The most common causes of secondary amenorrhoea are pregnancy and the menopause.

Any condition in which there is a failure of ovulation will cause infertility, although many women with regular ovulation remain infertile. In some cases of infertility there are anatomical disorders, while in others it has been shown that the cervical mucus is 'hostile' to sperm, thus impeding sperm transport, or that the woman is producing antisperm antibodies. Another cause of infertility is luteal insufficiency. In this condition the corpus luteum fails to produce sufficient progesterone to maintain the pregnancy until progesterone synthesis is taken over by the foetoplacental unit. Treatment is by supplementation with a progestogen during the latter phase of the menstrual cycle until the foetoplacental unit develops; non-androgenic progestogens such as dydrogesterone are used to overcome the risks of virilization of a female foetus.

## 11.10 Therapeutic uses and adverse effects of oestrogenic and progestogenic steroids

Probably the most common uses of oestrogens and progestogens are as oral contraceptives or hormone replacement therapy either postmenopausally or in cases of ovarian insufficiency; these indications are covered elsewhere in this chapter. Other uses for these hormones, and drugs which affect their activity, include fertility control and the treatment of certain hormone dependent cancers.

High doses of oestrogens can be used as a postcoital contraceptive. If oestrogens are administered within 72 hours of unprotected coitus the endometrial environment alters so that implantation cannot occur. A similar approach is to use a progesterone receptor antagonist such as mefipristone. This agent reduces the supportive effect of progesterone on the endometrium and therefore induces menstruation. An important difference between the postcoital oestrogen contraceptive and the progesterone antagonist is that the former prevents implantation, and therefore prevents pregnancy, while the latter causes failure of an established pregnancy.

Oestrogen receptor antagonists can also be used in the treatment of some forms of breast cancer. Approximately 40 per cent of breast tumours are oestrogen dependent and tumour regression can therefore be induced either by ovariectomy or by use of oestrogen antagonists such as tamoxifen or clomiphene. Oestrogens can also be used in males for the treatment of prostatic carcinoma, an androgen dependent tumour (see Chapter 10). Progestogens can be used in the treatment of endometriosis, a condition in which endometrial cells proliferate at sites outside the uterus, and in endometrial cancer.

Side-effects of oestrogens and progesterone range from the severe, such as breast cancer and cardiovascular disease, to the more minor. Details of the more severe risks are covered in the sections covering oral contraceptives and postmenopausal hormone replacement therapy. The more minor side-effects of oestrogens include nausea and vomiting, fluid retention, headache, breast enlargement and discomfort, weight gain and mood changes. Adverse effects of progestogens include amenorrhoea, fluid retention, weight gain and increased growth of body hair.

## 11.11 Oral and other hormonal contraceptives

The normal stimulus for ovulation is the mid-cycle surge in LH secretion, thus any treatment that reduces the LH surge is likely to prevent ovulation. Under normal conditions progesterone reduces the secretion of LH by negative feedback, thus administration of exogenous progesterone would be expected to prevent ovulation. Early attempts to test this hypothesis used oral administration of natural progesterone; however because of the high degree of first pass metabolism of this steroid very large doses were required. The major advance in the development of an acceptable oral contraceptive came with the synthesis of orally active progestogens such as norethisterone (known as norethindrone in the USA). Using these synthetic progestogens, which are derived from androgenic steroids, it was shown that it was possible to induce effective, reversible infertility. At the low doses of progestogen used it was found that ovulation was occurring in up to 60 per cent of the menstrual cycles studied, but the incidence of pregnancy remained low. It is now known that progestogens act to prevent pregnancy by at least three separate mechanisms: the prevention of ovulation, which may be only partial; an increase in the viscosity of cervical mucus, which prevents sperm transport; and the development of an endometrium into which a fertilized ovum is unable to implant. It is the combination of all of these factors that results in effective contraception.

The failure rate of progesterone oral contraceptives is typically three pregnancies per 100 women/years.

The progestogen-only oral contraceptive is taken daily, continuously. The contraceptive efficacy of the treatment is such that there are approximately three pregnancies each year for each 100 users (hundred women years, HWY). This poor contraceptive efficacy is partially due to the fact that the doses of progestogen used do not reliably inhibit ovulation and partially due to the fact that the drug must be taken at the same time each day; it has been suggested that the duration of effect of a single dose of progestogen is barely 24 hours, thus missing the next dose by as little as three hours can markedly reduce the contraceptive efficacy. A more recent advance has been the administration of the progestogens by intramuscular injection or subcutaneous implantation. These methods provide six weeks to five years of contraception from a single administration, and because they overcome the need to take a 'pill' regularly there are no pregnancies associated with delayed dosing. The most commonly reported problem of progestogen-only contraceptives is irregular, heavy menstruation. The major advantage of these types of contraceptives is their acceptability for use in women who cannot tolerate oestrogens because of an existing underlying condition such as a blood clotting disorder. There have also been attempts at targeted delivery of low doses of progestogens to the reproductive organs. In one development the progestogen is released from a silicon rubber ring that is inserted into the vagina; the extent of drug absorption is sufficient to produce contraception but the effects of the drugs at other sites around the body are reduced. The ring is removed for seven days of each cycle in order to allow a 'withdrawal bleed' to occur, which overcomes the problems of the unpredictable menstruation, but problems with user acceptability and vaginal abrasion have hampered attempts to introduce the device to the market.

Following the early trials of the progestogen-only contraceptive an oestrogen was added to the treatment regime in an attempt to control the irregular, heavy menstruation. This addition of an oestrogen gave rise to the first combined oral contraceptive. Combined oral contraceptives are normally administered daily for 21 days, followed by a seven-day drug free period. As with the progestogen-only treatment, the combined oral contraceptives reduce fertility by inducing a thickening of cervical mucus and changes in the endometrium but they are also more effective inhibitors of ovulation. During the seven drug-free days the withdrawal of the progestogenic support causes endometrial atrophy and therefore a menstrual-like bleed, but contraceptive cover is maintained. The combined oral contraceptives are much more effective than progestogen-only contraceptives, with a theoretical failure rate of approximately three pregnancies per 10 000 women per year (0.03 pregnancies per HWY, see later); the regular, predictable 'menstruation' improves user acceptability and provides reliable reassurance that pregnancy has not occurred, but the addition of the oestrogen component increases the risk of adverse effects (see later).

In an attempt to reduce the incidence and severity of the adverse effects, lower dose combined oral contraceptives have been introduced. Another approach to decreasing the total monthly dose of drug is to use sequential or phasic combined oral contraceptives (see Table 11.2). With these the doses of each of the components are varied throughout the 21 days of treatment; for example the dose of oestrogen may be kept constant while the dose of progestogen increases during the middle phase, or the dose of oestrogen may increase for the middle seven days of treatment while the progestogen increases step-wise throughout the whole 21 days. In addition to the lower total monthly doses, these treatment regimens are believed to mimic the natural menstrual cycle more closely than the single dose (monophasic) combined oral contraceptives, and therefore less likely to cause adverse effects (see later).

> The failure rate of combined oral contraceptives is three pregnancies per 1000 women per year.

In controlled clinical trials the efficacy of the combined oral contraceptives is such that there are only about three pregnancies per 10 000 women per year (0.03 pregnancies per HWY); however in general use in the developed world the failure rate is nearer 30 pregnancies per 10 000 women per year (0.3 pregnancies per HWY), and worldwide the failure rate approaches eight pregnancies per HWY. This indicates that the drugs themselves are very effective contraceptives, but there can be problems with how they are used. The reason for the differences in failure rate is that in clinical trials the users are monitored closely and the medication is taken correctly. In some less developed populations, however, the users do not understand the requirement to follow the manufacturers' instructions. In order for the drugs to be effective they must be taken consistently at the correct times. If a user fails to take the medication, or is late taking it, some contraceptive efficacy will be lost. The same may occur if the user suffers from vomiting or diarrhoea shortly after taking the 'pill' so that the drugs are not adequately absorbed, or if the user is suffering from some other form of condition or taking some other medication which may interfere with the actions of the contraceptives. These user dependent risks of contraceptive failure are greater in progestogen-only oral contraceptives and the very low dose (phasic) combined contraceptives because these regimes are less tolerant of errors.

Since their introduction in the early 1960s, the oral contraceptives have been associated with a range of adverse effects which range from the relatively minor such as chloasma (freckles) to the severe such as breast cancer. Among the less troublesome side-effects are weight gain, breast tenderness and loss of libido, all of which can usually be resolved by changing to a brand of contra-

**Table 11.2** Available forms of oral contraceptives (reproduced, with permission of the publisher, from the *Monthly Index of Medical Specialities (MIMS)*, in which details of available preparations are updated monthly)

## ORAL CONTRACEPTIVES

| Pill Type | Preparation | Manufacturer | Oestrogen (mcg) | Progestogen (mg) | |
|---|---|---|---|---|---|
| **Combined** | | | | | |
| *Ethinyloestradiol/* | Loestrin 20 | Parke-Davis | 20 | 1 | *norethisterone acetate* |
| *norethisterone type* | Loestrin 30 | Parke-Davis | 30 | 1·5 | *norethisterone acetate* |
| | Brevinor | Searle | 35 | 0·5 | *norethisterone* |
| | Ovysmen | Janssen-Cilag | 35 | 0·5 | *norethisterone* |
| | Norimin | Searle | 35 | 1 | *norethisterone* |
| *Ethinyloestradiol/* | Microgynon 30 (also ED) | Schering HC | 30 | 0·15 | |
| *levonorgestrel* | Ovranette | Wyeth | 30 | 0·15 | |
| | Eugynon 30 | Schering HC | 30 | 0·25 | |
| | Ovran 30 | Wyeth | 30 | 0·25 | |
| | Ovran | Wyeth | 50 | 0·25 | |
| *Ethinyloestradiol/* | Mercilon | Organon | 20 | 0·15 | |
| *desogestrel* | Marvelon | Organon | 30 | 0·15 | |
| *Ethinyloestradiol/* | Femodene (also ED) | Schering HC | 30 | 0·075 | |
| *gestodene* | Minulet | Wyeth | 30 | 0·075 | |
| *Ethinyloestradiol/* | Cilest | Janssen-Cilag | 35 | 0·25 | |
| *norgestimate* | | | | | |
| *Mestranol/* | Norinyl-1 | Searle | 50 | 1 | |
| *norethisterone* | Ortho-Novin 1/50 | Janssen-Cilag | 50 | 1 | |
| **Biphasic & Triphasic** | | | | | |
| *Ethinyloestradiol/* | BiNovum | Janssen-Cilag | 35 | 0·5 | (7 tabs) |
| *norethisterone* | | | 35 | 1 | (14 tabs) |
| | Synphase | Searle | 35 | 0·5 | (7 tabs) |
| | | | 35 | 1 | (9 tabs) |
| | | | 35 | 0·5 | (5 tabs) |
| | TriNovum | Janssen-Cilag | 35 | 0·5 | (7 tabs) |
| | | | 35 | 0·75 | (7 tabs) |
| | | | 35 | 1 | (7 tabs) |
| *Ethinyloestradiol/* | Logynon (also ED) | Schering HC | 30 | 0·05 | (6 tabs) |
| *levonorgestrel* | | | 40 | 0·075 | (5 tabs) |
| | | | 30 | 0·125 | (10 tabs) |
| | Trinordiol | Wyeth | 30 | 0·05 | (6 tabs) |
| | | | 40 | 0·075 | (5 tabs) |
| | | | 30 | 0·125 | (10 tabs) |
| *Ethinyloestradiol/* | Tri-Minulet | Wyeth | 30 | 0·05 | (6 tabs) |
| *gestodene* | | | 40 | 0·07 | (5 tabs) |
| | | | 30 | 0·1 | (10 tabs) |
| | Triadene | Schering HC | 30 | 0·05 | (6 tabs) |
| | | | 40 | 0·07 | (5 tabs) |
| | | | 30 | 0·1 | (10 tabs) |
| **Progestogen only** | | | | | |
| *Norethisterone type* | Micronor | Janssen-Cilag | — | 0·35 | *norethisterone* |
| | Noriday | Searle | — | 0·35 | *norethisterone* |
| | Femulen | Searle | — | 0·5 | *ethynodiol diacetate** |
| *Levonorgestrel* | Microval | Wyeth | — | 0·03 | |
| | Norgeston | Schering HC | — | 0·03 | |
| | Neogest | Schering HC | — | 0·075 | *norgestrel* |

*Converted (>90%) to norethisterone as the active metabolite

ceptive with different doses or ratios of oestrogen and progestogen. The most severe of the reported adverse effects, cervical cancer, breast cancer and thrombosis, require greater consideration.

### 11.11.1   *Oral contraceptives and thrombosis*

Thrombosis has been associated with use of combined oral contraceptives since their introduction. The oestrogen component of the 'pill' increases the production of various clotting factors and therefore decreases clotting time; progestogens do not affect clotting factors but they do increase plasma concentrations of high density lipoproteins, which is associated with an increased risk of cardiovascular disease. Since the introduction of combined oral contraceptives the dose of the oestrogenic component has decreased from typically 1 mg of ethinyloestradiol in 1965 to 0.7 mg per month in 1996. Over the same period the monthly dose of the progestogen norethisterone has decreased from 42 mg to 16 mg. The risk of thrombosis and cardiovascular disease has thus been reduced to almost baseline levels by the introduction of the low dose preparations, although it is recommended that oral contraceptives should be used with caution in those women with a history of cardiovascular disease or who are predisposed to cardiovascular disease, for example smokers. In 1995, however, results of a large scale study again linked oral contraceptive use with an increased risk of venous thromboembolism (pulmonary embolism and deep vein thrombosis). This increased risk was associated with the newly introduced, increased potency, progestogens gestodene and desogestrel. The incidence of thromboembolism in healthy young women is approximately five cases per 100 000 women per year; in users of 'standard' combined oral contraceptives there is a threefold increase in risk, but in users of the preparations containing either of the two new progestogens the incidence of thromboembolism was 30 cases per 100 000 women years. These figures may indicate that use of the newer progestogens is associated with an increased risk of thromboembolism, but it has also been suggested that these preparations were routinely used in women with higher risk of thromboembolism because of the original belief that they were safer than the standard 'pills', and therefore a higher incidence of thromboembolism may have been expected. For comparison, there are 45 cases of thromboembolism per 100 000 pregnancies.

### 11.11.2   *Oral contraceptives and breast cancer*

Several studies have suggested that use of oral contraceptives may increase the risk of breast cancer. For example, one study reported that use of oral contraceptives for four to eight years increased the risk of developing breast cancer before the age of 36 by 43 per cent and that more than eight years of use increased the risk of breast cancer by 74 per cent. It was however noted that those contraceptives containing less than 50 $\mu$g of oestrogen per day had lower relative risks, and that progestogen-only contraceptives actually reduced the risk by 16 per cent. A more recent study suggests that these figures are misleading. In the report of this later study it was suggested that the use of combined oral contraceptives accelerates the presentation of breast cancer; thus the appearance of breast cancer in young women is greater among 'pill' users, but does not increase the overall prevalence of breast cancer. It is therefore believed that combined oral contraceptives promote the growth of existing breast tumours but do not actually cause the disease.

### 11.11.3  *Oral contraceptives and cervical cancer*

It is sometimes suggested that the risk of cervical cancer is related to the sexual activity of the woman. In 1983, however, it was reported that the use of combined oral contraceptives containing 50 $\mu$g of oestrogen increased the risk of cervical cancer. The risk was elevated by 60 per cent in women who had used the 'pill' for four to six years and was more than doubled in women who had used it for more than six years; there were no such increases in risk among users of intrauterine contraceptive devices, nor was there any evidence of any differences in sexual activity between the users of the different forms of contraceptive. It is now recognized that combined oral contraceptives increase the incidence of cervical cancer, a treatable condition, but that they also decrease the incidence of the more serious uterine and ovarian tumours.

It is much safer to take the 'pill' for one year than to be pregnant for nine months.

As a final comment, when the relative risks of oral contraceptive usage are considered they must be compared with the risks involved with other activities. As mentioned above, the risks of thromboembolism are the same for four years of 'pill' use as they are for one year of pregnancy, similarly, worldwide, there are 200 times more deaths during childbirth than there are deaths caused by oral contraceptives.

## Summary

- In the absence of androgenic influence, a foetus will develop female genitalia, irrespective of genetic gender; oestrogens do not influence foetal development. In late childhood the secretion of oestrogens and progesterone by the cells of the ovary initiates the changes characteristic of female puberty. These changes include breast development, deposition of subcutaneous fat, long bone growth, growth of female pattern pubic hair and the initiation of ovulation.

- Ovulation occurs as part of a cycle of events (the menstrual cycle) controlled by the secretion of the hypothalamic hormone gonadotrophin releasing hormone, and the pituitary hormones follicle stimulating hormone and luteinizing hormone. Other changes that occur as part of this cycle are the maturation of a Graafian follicle, the thickening and subsequent atrophy of the endometrium and changes in cervical dilatation and cervical mucus viscosity. All of these changes are concerned with the enhancement of fertility.

- A deficiency of female sex hormones may cause a delay in puberty, and a failure to undergo the normal cyclical changes in reproductive physiology. Sex hormone secretion normally ceases in later life (the menopause), resulting in infertility and some regression of the secondary sexual characteristics. Excess sex hormone secretion may induce precocious puberty.

- Female sex hormones may be used therapeutically in the treatment of hormone deficiency, or for hormone replacement therapy after the menopause. They may also be used to reduce fertility (contraception), but their use is not without risk of side-effects.

## Selected reading

Binkley, S.A., 1995, Estrogens and progesterone and the ovary. In Binkley, S.A., *Endocrinology*, New York: HarperCollins Publishers, 364–404

Fritsch, M.K., 1994, Estrogens, progestins, and oral contraceptives. In Brody, T.M., Larner, J. and Neu, H.C. (eds), *Human Pharmacology: Molecular to Clinical*, 2nd edn, St Louis: Mosby, 482–500

Hadley, M.E., 1992, Endocrinology of pregnancy, parturition, and lactation. In Hadley, M.E., *Endocrinology*, 3rd edn, Englewood Cliffs: Prentice-Hall, 505–530

Laycock, J. and Wise, P., 1996, Female reproductive endocrinology. In Laycock, J. and Wise, P., *Essential Endocrinology*, 3rd edn, Oxford: Oxford University Press, 163–202

Williams, C.L. and Stancel, G.M., 1996, Estrogens and progestins. In Hardman, J.G., Linbird, L.E. and Gilman, A.G. (eds), *Goodman and Gilman's The Pharmacological Basis of Therapeutics*, 9th edn, New York: McGraw-Hill, 1411–1440

# 12 Hormonal Control of Growth and Development

## 12.1 Introduction

As stated in the opening chapters, the endocrine system provides a means by which processes that occur at multiple sites around the body can be controlled in a continuous, long-lasting manner. A prime example of such endocrine influence is the control of growth. Growth may be simply defined as an increase in size, but there are several processes which may be involved. Growth may result from an increase in the number of cells within a particular tissue, or from an enlargement of each of the individual cells. Growth may also be a result of an increased production of extracellular substances. With very few exceptions, nearly all hormones have the ability to stimulate growth in their target tissue; it may therefore seem strange that such an important a process as growth, which begins immediately after conception and continues, at least to some extent, until the time of death, should not be discussed until the final chapter of a textbook of endocrinology, but to discuss it earlier presents almost insurmountable difficulties. In order to appreciate the manner in which the actions of a wide range of different hormones are integrated in the control of growth and development one must first understand the physiology of each individual hormone. Such physiology has been presented in the preceding chapters; the aim of this chapter is to illustrate how many different hormones can act in an orchestrated way to bring about normal growth, and how a disorder of just one of the endocrine components involved can be manifested as a growth abnormality.

Non-endocrine factors such as malnutrition and deprivation may also retard growth.

When considering growth it is important to remember that many factors produce an influence. Diminished growth, for example, may be caused by malnutrition, malabsorption, chronic hypoxia due to congenital heart disease, chronic renal disease, asthma, prolonged steroid therapy, hypothyroidism, hypoparathyroidism, precocious puberty, psychological and social deprivation. Similarly there are great ethnic variations, and genetic influences. These factors must all be assimilated when trying to assess the importance of endocrine disorders in the aetiology of growth abnormalities (see Figure 12.1). It is in the light of these factors that the effects of various endocrine influences on growth will be described.

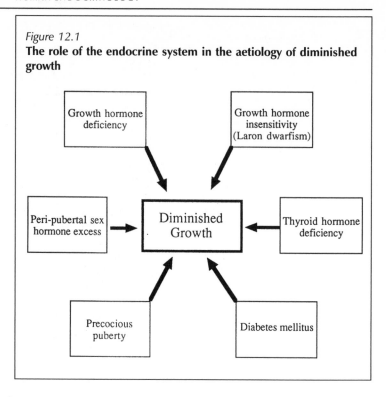

Figure 12.1
**The role of the endocrine system in the aetiology of diminished growth**

## 12.2 Growth hormone

Most effects of growth
hormone are mediated
by the somatomedins.

Probably the most important hormone in the control of linear
growth is growth hormone (GH). This hormone is secreted by the
somatotrophe cells of the anterior pituitary gland under the control
of growth hormone-releasing hormone (GH-RH), and to a lesser
extent, somatostatin, from the hypothalamus (see Chapter 2).
Within the circulation GH is transported bound to a plasma protein,
growth hormone binding protein, and it is thus only the unbound
portion (30 per cent) which is biologically active. The amino acid
sequence of the growth hormone binding protein resembles that of
the extracellular portion of the GH receptor. The GH receptor has
been shown to be membrane bound and composed of two identical,
membrane spanning protein chains which are structurally similar to
those of the prolactin receptor (Chapter 2, Figure 2.4). The second
messenger process for GH is poorly defined, but in some tissues the
tyrosine kinase activity of the receptor proteins results in phosphorylation of intracellular proteins resulting in gene expression and
protein synthesis, while in other cells the receptor protein is linked
to a G-protein, the activation of which results in the production of
the second messengers inositol triphosphate and diacyl glycerol.
The primary target tissue for GH is the liver where it stimulates
the production and release of the somatomedins. Like GH, the
somatomedins are transported within the plasma extensively (95
per cent) bound to plasma proteins. In many tissues the actions

of GH on growth have been shown to be mediated by these somatomedins.

The two somatomedins that have been isolated and characterized fully have been found to have amino acid sequences that closely resemble that of insulin; they have thus been labelled insulin-like growth factors I and II (IGF-I and II). Whereas insulin is a polypeptide with 51 amino acids arranged as an $\alpha$-chain of 21 amino acids and a $\beta$-chain of 30 amino acids linked by disulphide bridges, IGF-I and II are single peptide chains of 70 and 67 amino acids respectively; in adult humans IGF-II is three times more abundant than IGF-I. Specific receptors for IGF-I and II have been identified: IGF type 1 receptors which have high affinity for IGF-I, lower affinity for IGF-II and low affinity for insulin; and IGF type 2 receptors which show greatest affinity for IGF-II, lesser affinity for IGF-I and no affinity for insulin. At high concentrations, both IGF-I and II are able to stimulate insulin receptors. Like the insulin receptor, the IGF receptors are membrane bound tyrosine kinase linked receptors, the stimulation of which not only results in phosphorylation of intracellular proteins which increase cellular differentiation and replication, but also phosphorylation of the receptor protein itself which produces an amplification of the receptor mediated effects.

The predominant physiological effect of GH is the promotion of linear growth that occurs during adolescence. This effect results from an increase in protein synthesis and extracellular collagen deposition, which is mediated at the cell nucleus. Growth hormone also promotes the availability of the nutrients and elements required for growth by stimulating amino acid uptake across the cell membrane, increasing uptake of calcium by the gut, increasing uptake of fatty acids by muscle and promoting lipolysis. With the exception of the latter, all of these effects involve an action at the level of the cell membrane. All of the above mentioned actions of GH are probably mediated by the somatomedins which may be produced either in the liver or within the target tissue. There are also direct effects of the interaction of GH with its receptor; the most obvious of these is the stimulation of the synthesis of the IGFs, but expression of IGF receptors may also be induced by the actions of GH.

In high concentrations, GH has effects which are opposite to those seen at lower levels. Thus the actions of normal concentrations of GH are generally mediated by the somatomedins and tend to complement the actions of insulin in that there is increased cellular uptake of fatty acids and amino acids and enhanced protein synthesis. This should not be surprising, given the structural similarities of insulin and IGF-I and II. At high concentrations of GH, however, there is an increase in blood glucose because of the stimulation of lipolysis. This is seen as a diabetogenic effect of GH and is believed to be a direct consequence of the stimulation of the GH receptor, rather than an effect mediated by the somatomedins.

Growth hormone is responsible for the control of growth that occurs at the various stages of life; it is also concerned with the control of tissue turnover and repair. Foetal growth, however,

Foetal growth differs
from growth that
occurs later in life in
that it is independent
of growth hormone.

differs from the growth that occurs later in life in that it appears to
be relatively independent of growth hormone. In the foetus and in
very young children the circulating concentrations of IGF-I and II
are very low. It has thus been suggested that growth during this
stage is controlled by an embryonic form of somatomedin which
is produced by all foetal cells, without there being a necessity for
stimulation by GH. Embryonic somatomedin is present in high
concentrations in early life, but declines in adult life. In adults
the secretion of growth hormone and the production of the somatomedins is stimulated by a range of physiological stimuli including
hypoglycaemia, decreased plasma fatty acid concentrations and
increased plasma amino acid concentrations. Growth hormone secretion is also stimulated during stress. Both growth hormone itself,
and the somatomedins are able to exert negative feedback on growth
hormone secretion by effects either on the hypothalamus or on the
somatatotrophes of the pituitary gland. Growth hormone secretion
is stimulated by oestrogens, probably by an alteration in the number
of receptors for the hypothalamic hormones on the surface of the
somatotrophes. It may be this action of the oestrogens that causes
the spurt in linear growth that normally occurs at puberty; at the
time of puberty the increase in secretion of the male and female sex
steroids is accompanied by an increase in growth hormone secretion which leads to a marked acceleration in the rate of growth. This
continues for four to five years before growth ceases when the sex
hormones cause fusion of the epiphyses.

### 12.2.1 Growth hormone deficiency

Growth hormone deficiency is rare, but when present in children it
results in dwarfism. In such cases, the intrauterine and early growth
is normal, but growth rapidly falls behind that of peers after the age
of two years. In 50 per cent of growth hormone deficient dwarfs
there are also deficiencies of other hormones resulting in Graves'
disease (Chapter 4), Addison's disease (Chapter 9) or diabetes
mellitus (Chapter 7). This corelationship between the disorders
suggests a single underlying disorder, probably autoimmune in nature. In some cases the growth deficiency occurs because of an
inability to produce the somatomedins; the concentrations of circulating growth hormone are in the normal range. This form of dwarfism is known as Laron dwarfism. Interestingly, another feature of
Laron dwarfism is the absence of the high affinity growth hormone
binding protein in the plasma, suggesting that the underlying feature of the disorder is the lack of growth hormone receptors. Some
ethnic groups are unable to synthesize IGF-I, although IGF-II syn-

The pygmies of Africa
are short because
of a genetic inability
to produce
somatomedins.

thesis is normal. In these groups, for example the African pygmies,
the Loja of Equador and the Mountain Ok of New Guinea, the
genetic variant results in small stature populations.

In adults, damage to the anterior pituitary gland may result in a
cessation of secretion of several hormones, including growth hormone. In these individuals, however, because normal stature has

already been achieved, the growth hormone deficiency has traditionally been seen to be of little importance (see later).

Treatment of growth hormone deficiency is by hormone replacement therapy. Because of the specificity of the human growth hormone receptor, however, it is not possible to use growth hormone of animal origin. This meant that early therapies used hormonal material extracted from human cadavers. Recent fears concerning the transmission of neurological disorders such as Creutzfeldt-Jakob disease have now rendered such sources unacceptable. Recent developments have thus concentrated on the production of human growth hormone using recombinant DNA techniques; however the resultant products are very expensive. Replacement therapy with these hormones, if initiated before the closure of the epiphyses, may result in extra growth, but many authorities question the cost effectiveness of the treatment as the growth achieved may be as little as an extra four centimetres.

Growth hormone treatment has also recently been shown to be useful in adult victims of anterior pituitary damage where feelings of fatigue, which were previously refractory to treatment by replacement of the other pituitary hormones, have been seen to be relieved. Growth hormone has also been shown to be of use in the treatment of some symptoms of acquired immune deficiency syndrome (AIDS) (see Box 12.1).

### 12.2.2  *Growth hormone excess*

In cases of hypersecretion of growth hormone there is excessive growth, the most common cause of which is a secretory tumour of the anterior pituitary gland although use of $\beta$-adrenoceptor antagonists has also been associated with increased growth hormone

---

Box 12.1  **The use of growth hormone in the treatment of weight loss in AIDS patients**

One-hundred-and-seventy-eight patients suffering from AIDS associated wasting received injections of either human growth hormone or placebo, daily for three months. Patients underwent a treadmill test to assess physical performance and were weighed both at the beginning and at the end of the trial. By the end of the three months, those patients receiving growth hormone had a mean increase in body weight of 1.6 kg despite the fact that there had been a loss of fat: the mean increase in lean body weight was 3.0 kg. In contrast, those patients receiving placebo injections neither gained nor lost body weight, and there was no change in their fat-to-lean ratio.

When the results of the treadmill test were considered, it was found that the patients receiving placebo had increased their performance by a mean of 2 per cent over the three month period, while those receiving growth hormone had improved by a mean of 13 per cent.

There were unfortunately some side effects of the growth hormone treatment, the most common being oedema with joint and muscle pain and diarrhoea. The incidence of side-effects was reduced by a decrease of growth hormone dose.

Sadly, the growth hormone treatment had no effect on the survival rate of these patients.

Schambelan, M. (1996) *Annals of Internal Medicine* **125** 873

secretion. Excess secretion of growth hormone occurs in approximately 50 individuals per million of population, less than 5 per cent of which occur before puberty. Growth hormone excess before puberty results in gigantism, in which case the symptoms include excessive growth, to a height in excess of 2.5 metres, and increased muscle mass (see Figure 12.2). More commonly the condition presents in adults after normal closure of the epiphyses, when it is characterized by a thickening of the skin around the face leading to a coarsening of the features and increased growth of the nose and ears. There is also growth of the bones of the skull which results in the development of a jutting jaw and a prominent forehead; the hands and feet also become enlarged. This condition, which is called acromegaly, may take 15–20 years to develop, thus the changes in appearance may be missed by close friends and relatives; infrequent acquaintances are more likely to notice the progression of the disorder. Other symptoms of the condition, such as visual disturbances and other endocrine disorders, are caused by compression of surrounding structures, for example the optic nerve, by the developing tumour. The excessive secretion of growth hormone also results in continued growth of the internal organs thus there is often enlargement of the spleen, liver and heart. If left untreated, the condition often results in death due to the effects on the cardiovascular system, for example atherosclerosis and cardiomyopathy.

Most cases of acromegaly, and gigantism, can be treated with drug therapy. Older therapies used dopamine agonists to stimulate

> Acromegaly is characterized by growth of bones other than the long bones of the limbs.

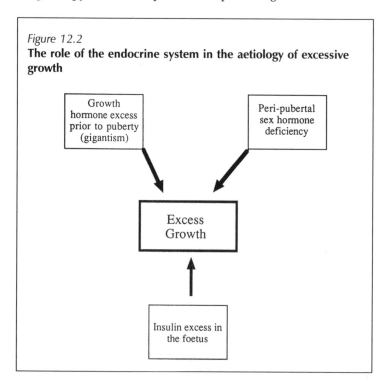

Figure 12.2
**The role of the endocrine system in the aetiology of excessive growth**

the neuronal pathways that are known to reduce growth hormone secretion; this was succesful in about 70 per cent of cases. More recently it has been posible to administer somatostatin analogues such as octreotide (proprietary name Sandostatin) to bring about reduced secretion. In cases where pharmacotherapy is unsuccessful the tumour is usually removed or destroyed either by surgery or by radiotherapy.

## 12.3 Sex hormones

As mentioned above, another group of hormones that have profound influences on growth and development are the sex hormones. The anatomical effects of the sex hormones begin before birth where the presence of the male sex hormones results in the development of male internal and external genitalia, while absence of the androgens results in development of female anatomical features. Following birth the normal growth of the infant is independent of the sex hormones until puberty. At the time of puberty there is initiation of the secretion of oestrogens and progesterone in females and the initiation of testosterone secretion in males. These hormones cause the development of the normal female and male secondary sexual characteristics. They also act together with growth hormone to cause the spurt in linear growth that occurs at puberty. As described in Chapters 10 and 11, in males the normal increase in secretion of testosterone at puberty results in an increase of height of approximately 19 cm between the ages of 13 and 16; this compares with an increase of only 15 cm between the ages of 7 and 10 when growth hormone is acting in the absence of testosterone. Comparable figures for females, who exhibit the growth spurt earlier than males are 19 cm between the ages of 10 and 13 compared with 15 cm between the ages of 7 and 10. Importantly it is also the sex hormones that cause the closure of the epiphyses and the cessation of further long bone growth at the age of 18 in males and 15 in females.

It is not surprising, therefore, that disorders of sex hormone secretion cause abnormalities of growth. In the case of hypogonadism, there is a lack of the pubertal growth spurt, thus the individuals are shorter than their peers; however the sex hormone deficiency also means that the epiphyses do not close in late adolescence, which means that the effects of growth hormone on bone growth continue beyond the age of 20. It is thus very common for sufferers of hypogonadism to achieve a greater height than average (see Figure 12.2). In the case of hypergonadism the opposite pattern of growth occurs. Initially the linear growth spurt exceeds that of peers, thus in early teenage years the individual is taller than average, but the early closure of the epiphyses, and the limitation of further growth means that by the age of 17 the individual is shorter than average.

Hypergonadism causes short, stocky stature.

The treatment of growth disorders related to abnormalities of sex hormone secretion is normally depedent upon the underlying cause

of the disorder. Thus in cases of hypogonadism due to enzyme deficiency, for example congenital adrenal hyperplasia, sex hormone
replacement therapy may be appropriate. In Turner's syndrome, a
disorder in which a deficiency of an X chromosome results in oestrogen deficiency, it has recently been shown that administration of
growth hormone is effective in treating the growth deficit.

In cases of hypergonadism due to the presence of a secretory
tumour, the normal course of action is the removal of the source
of the excess hormones, although the treatment has no effect on
growth if closure of the epiphyses has already occurred.

## 12.4 Thyroid hormones

Thyroid hormones also influence growth and development. In the
case of hyperthyroidism the increased basal metabolic rate results
in utilization of skeletal proteins for metabolism, culminating in
growth retardation.

Thyroid deficient
dwarfism is
characterized by
retention of juvenile
proportions; in growth
hormone deficient
dwarfism normal
maturation takes place.

Growth retardation is also present in hypothyroidism, but due to
a different mechanism. Thyroid hormones are important in the
control of normal development, thus a deficiency normally causes
a delay in the onset of puberty. Because of this, in children with
thyroid hormone deficiency there is a failure to undergo the spurt of
growth that normally occurs at puberty; they therefore retain juvenile proportions with short limb length compared to trunk length
and their faces also retain an immature appearance. These factors
are important for the differentiation of thyroid deficient dwarfism
and growth hormone deficient dwarfism, where there is a failure of
growth, but normal signs of maturation, for example elongation of
the limbs. Treatment of hypothyroid dwarfism is by administration
of thyroid hormones. In most cases such treatment is sufficient to
enable achievement of normal height and sexual maturation.

## 12.5 Insulin

Insulin plays an important role in the control of carbohydrate
metabolism and therefore the availability of nutrients, but it may
also have effects on growth that are independent of nutrition. Such
effects may be related to the similarity between the structure of
insulin and those of the somatomedins. Children with insulin
deficiency, as in insulin dependent diabetes mellitus, fail to grow
despite the presence of normal concentrations of growth hormone.
Similarly, Laron dwarfs in whom there is a deficiency of IGF production also tend to be deficient in insulin. This suggests a relationship between the actions of growth hormone, via the somatomedins,
and the effects of insulin. It is also known that insulin itself can
cause growth, for example in cases of excesive insulin secretion. In
non-insulin dependent diabetes mellitus for example, there is often
enlargement of the kidneys and adrenal glands and overgrowth of
the extremities. This effect is probably due to an action of insulin on

IGF receptors. Children of diabetic mothers, in whom the foetal pancreas may have been over-secreting insulin in order to overcome the maternal deficit, are also frequently large. It therefore appears that normal growth may also be dependent on normal secretion of insulin, although the precise role of insulin in the control of growth remains unclear.

## 12.6 Conclusion

The control of growth and maturation is very complex, and thus involves the interplay of many separate endocrine glands and hormones. It is because of this that many endocrine disorders, if left untreated, may result in abnormalities of growth. It is also because of the multiplicity of control that it is often very difficult to ascertain the single endocrine dysfunction responsible for any given incidence of growth retardation.

## Summary

- Growth and tissue repair is a life-long process that occurs throughout the body. The endocrine system is one of the most important systems in its control and therefore abnormal secretion of only one of a number of hormones can cause abnormal growth. There are, however, many other, non-endocrine factors that can cause growth retardation, for example malnutrion and psychological deprivation.

- The most important hormone in the control of linear growth is growth hormone from the anterior pituitary gland. Growth hormone acts on the liver to cause the production of somatomedins, for example insulin-like growth factors, which are responsible for the promotion of all processes concerned with growth. A deficiency of growth hormone in children results in dwarfism, although other aspects of maturation are normal. Excess growth hormone in children causes the rare condition of gigantism; more commonly however the hormone excess occurs after the closure of the epiphyses when it causes acromegaly, which is characterized by a thickening of the skin and an enlargement of the bones of the skull.

- Male and female sex hormones act together with growth hormone to cause the spurt in linear bone growth that occurs at puberty, followed by closure of the epiphyses. Hypogonadism results in a delay in the spurt of growth, but because of the delayed closure of the epiphyses growth may continue for longer. Conversely, hypergonadism results in an early growth spurt but premature closure of the epiphyses causes reduced stature. A delay of maturation and growth may also occur in hypothyroidism resulting not only in diminished stature but also retention of juvenile proportions.

- Insulin deficiency may also cause growth retardation, independent of its effects on carbohydrate metabolism and the resultant diabetes mellitus. This effect on growth is probably related to its structural similarities to the somatomedins.

# Selected reading

Binkley, S.A., 1995, Growth hormone from the anterior pituitary. In Binkley, S.A., *Endocrinology*, New York: HarperCollins Publishers, 107–126

Laycock, J. and Wise, P., 1996, Growth and development. In Laycock, J. and Wise, P., *Essential Endocrinology*, 3d edn, Oxford: Oxford University Press, 321–337

Woods, K.A., Weber, A. and Clarke, A.J.L., 1995, The molecular pathology of pituitary hormone deficiency and resistance. In Thakker, R.V. (ed.), *Baillière's Clinical Endocrinology and Metabolism, International Practice and Research. Vol. 9, No. 3, Genetic and Molecular Biological Aspects of Endocrine Disease*, London: Baillière Tindall, 453–487

# Appendix
# USA Drug Names

This appendix lists some of the cases where the US and UK proprietary names of drugs used to treat endocrine disorders, and hormone preparations, differ

| **Drug** | **UK name** | **US name** |
|---|---|---|
| ***Anterior pituitary gland*** | | |
| Leuprorelin | Prostap | Lupron |
| | | |
| ***Posterior pituitary gland*** | | |
| Vasopressin | Pitressin | Diapid |
| Desmopressin | Desmotabs | Concentraid |
| | Desmospray | Stimate |
| Demeclocycline | Ledermycin | Declomycin |
| Oxytocin | Syntocin | Syntocinon |
| | | |
| ***Gastrointestinal tract*** | | |
| Omeprazole | Losec | Prilosec |
| | | |
| ***Pancreas*** | | |
| Glibenclamide | Daonil | DiaBeta |
| | Euglucon | Glynase |
| | | Micronase |
| Glipizide | Glibenese | Glucotrol |
| | Minodiab | |
| Tolazamide | Tolanese | Tolinase |
| Tolbutamide | Rastinon | Orinase |
| | | |
| ***Adrenal medulla*** | | |
| Propanolol | Inderal | Betachron |
| Labetalol | Trandate | Normodyne |
| Metoprolol | Betaloc | Lopressor |
| | | Toprol XL |

### Adrenal cortex

| | | |
|---|---|---|
| Prednisolone | Precortisyl Forte | Articulose |
| | Prednesol | Delta-Cortef Predcor |
| Methylprednisolone | Medrone | Adlone Medralone M-prednisol |
| Triamcinolone | Kenalog Ledercort | Aristocort Azmacort Trilone Trymex |
| Betamethasone | Betnelan Betnovate Betnesol | Alphatrex Diprosone Psorion Valisone Uticort |
| Dexamethasone | Decadron | Decaderm Maxidex Solurex |
| Beclomethasone | Aerobec Becotide | Beclovent Vancenase Vanceril |
| Fluticasone | Flixotide Flixonase | Cutivate Flonase |
| Flunisolide | Syntaris | Aerobid Nasalide |

### Male reproductive system

| | | |
|---|---|---|
| Flutamide | Drogenil | Eulexin |

### Female reproductive system

| | | |
|---|---|---|
| Fosfestrol | Honvan | Stilphostrol |

### Oral contraceptives

| | | |
|---|---|---|
| Ethinyloestadiol 35 $\mu$g/ norethisterone 1 mg | Norimin | Norinyl Ortho-Novum |
| Ethinyloestradiol 35 $\mu$g/ norethisterone 0.5 mg | Brevinor Ovysmen | Brevicon Modicon Genora Nelona |
| Ethinyloestradiol 30 $\mu$g/ levonorgestrel 0.15 mg | Ovranette Microgynon | Levlen Nordette |
| Ethinyloestradiol 30 $\mu$g/ desogestrel 0.15 mg | Marvelon | Desogen Ortho-Cept |

# Index